IEE ELECTROMAGNETIC WAVES SERIES 7

SERIES EDITORS: PROFESSOR P. J. B. CLARRICOATS,
G. MILLINGTON,
E. D. R. SHEARMAN
AND J. R. WAIT

Reflector antenna analysis and design

Previous volumes in this series

Volume 1	Geometrical theory of diffraction for electromagnetic waves Graeme L. James
Volume 2	Electromagnetic waves and curved structures Leonard Lewin, David C. Chang and Edward F. Kuester
Volume 3	Microwave homodyne systems Ray J. King
Volume 4	Radio direction-finding P. J. D. Gething
Volume 5	ELF communications antennas Michael L. Burrows
Volume 6	Waveguide tapers, transitions and couplers F. Sporleder and H. -G. Unger

Reflector antenna analysis and design

P. J. Wood, B.Sc., Ph.D., C.Eng., M.I.E.E.
GEC–Marconi Electronics Ltd.
Great Baddow Research Laboratories
Great Baddow, Chelmsford
Essex, England

PETER PEREGRINUS LTD.
On behalf of the
Institution of Electrical Engineers

Published by: The Institution of Electrical Engineers, London and New York
Peter Peregrinus Ltd., Stevenage, UK, and New York

British Library Cataloguing in Publication Data

Wood, P J
Reflector antenna analysis and design. — (Institution of Electrical Engineers. IEE electromagnetic waves series; 7).
1. Antennas, Reflector
I. Title II. Series
621 .381'33 TK 7871.6 79-41447

ISBN 0 906048 21 4

Typeset at the Alden Press Oxford London and Northampton
Printed in England by A. Wheaton & Co., Ltd., Exeter

Preface

The various facets of the electrical design of a reflector antenna are usually best tackled by a judicious combination of

(*a*) empirical methods
(*b*) scalar aperture and Fourier transform theories
(*c*) vector electromagnetic models of the antenna diffraction processes.

Each of these alternative approaches has its own areas of strength and its own particular limitations and restrictions. In case (*c*), the limitations often result from the degree of complexity that is inevitably to be associated with the electromagnetic transformations in a practical reflector antenna. For example, an analytical technique may be posed in a form that appears promising as applied to an idealised scattering configuration such as a point source illuminating a paraboloid. However, it may prove difficult to generalise the algorithm to treat a more arbitrary (and realistic) antenna configuration. Alternatively, the drawback may be the computer time required to carry out a calculation.

This book places some emphasis on the induced current/spherical wave formulation of vector diffraction. It would, of course, be unfair to claim that this is a technique which is exempt from all restrictions of the type alluded to above. Nevertheless, the range of problems that can be treated in a numerically efficient manner by an essentially universal formulation in vector spherical waves is considerable. Thus, complete modelling of feed horn radiation properties via either theoretically derived or measured data, analysis of reflector scattering including far-field, near-field, focused or defocused antenna systems, and even pattern synthesis can come within its scope. It is hoped that this book will help draw attention to the expansion method, and to its potential value in many applications of interest to the microwave antenna engineer.

Much of the material in the chapters to follow has, directly or indirectly, come into being in response to particular requirements relating to tasks undertaken by the Microwave Antenna Group of Marconi Research Laboratories. It is a pleasure to acknowledge the encouragement and assistance given by my colleagues at MRL, without which this book would not be possible. Many engineers have contributed towards certain specific projects referred to briefly in the following chapters. Of my immediate colleagues, I would mention in particular J. R. Brain, N. J. L. Lockett, J. H. Cowan and R. W. Ashton.

Finally, I wish to thank Mrs. P. Atkins for typing much of the early draft, and my wife Sue for her perseverance in typing and proof reading the manuscript of this book.

P. J. Wood
Great Baddow Research Laboratories
Essex

Contents

of the feed system, which is such as to achieve a near optimum tradeoff between two principal loss contributions. These are the losses due to spill-over past the main and subreflectors, and due to nonuniform illumination of the main reflector aperture. The former can only be evaluated satisfactorily using diffraction techniques. Furthermore it is necessary to cater for the general case where the subreflector does not lie in the far-field region of the feed or feed system. A similar limitation applies for the main reflector in relation to the subreflector. A special type of synthesis technique has been developed to determine the optimum reflector profiles bearing in mind these requirements. This will be outlined in later chapters of this book.

Analytical techniques can also play a significant role whenever it is necessary to estimate the effect of small unintentional deviations from the nominal profile of the surface of a reflector. For example, the antennas employed in a tropospheric scatter communication system usually have parabolic section reflectors, offset-fed at their prime focus. Because of the nature of the communication mode, the off-axis pattern, as well as the boresight gain, contributes to the overall signal/noise ratio. One problem area involves the effect of high winds at the antenna installation. The first stage of an analytical investigation will be a structural analysis, taking into account wind loadings and mechanical constraints. The second stage is the electromagnetic scattering analysis, in which the complex surface deformation of the reflector is translated into gain and radiation pattern perturbations.

An investigation of this general type may also be carried out in other quite different contexts. For example, one of the most important sources of profile distortion for an antenna on board a satellite is thermal expansion. Solar illumination with partial shade results in uneven temperatures and consequent displacements in the antenna structure. In this case a mechanical analysis defines the reflector surface perturbations as a function of the angle of incidence of the sun's rays. The electromagnetic scattering analysis then yields the perturbed radiation pattern contours, and the degradation in signal strength over the coverage zone on the earth's surface.

The examples quoted have been chosen to illustrate how the antenna designer may sometimes need to place considerable emphasis on the use of numerical methods. In fact, an empirical approach would have substantial drawbacks for either of the two types of problem described above. Thus for the large Cassegrain antenna (perhaps 20 m diameter), the Rayleigh range may be 20 km or more. It may not then be possible to find a test range which would allow the secondary pattern performance to be measured directly, without substantial uncertainties owing to ground reflections. Alternatively, in the case of profile distortions, an empirical approach would demand the amalgamation of the facilities of a polar diagram test range for measuring the electrical properties, and of an environmental test chamber for inducing the distortions. This is rarely feasible.

In most subject areas the development of synthesis techniques tends to lag behind progress in analysis. Reflector antennas are no exception to this generalisation, and at the present time there are still many aspects of the synthesis problem which await investigation. Nevertheless, synthesis methods have been used in some of the earliest radar applications. The best known example is the 'cosec squared' radar surveillance antenna intended to provide a shaped elevation coverage to ensure a constant strength radar return from an overflying aircraft. In this case however, the associated synthesis technique was one which depended on geometrical optics concepts, and is therefore of somewhat limited application.

The reflector antenna represents an extremely versatile type of radiator, potentially capable of realising a close approximation to virtually any desired distribution of radiation. This includes synthesis of a pattern cut of given shape, or synthesis of a given shape of beam cross-section. The desired 'shaped' radiation characteristic can be achieved merely by contouring the profile of the reflecting surface. For high-frequency applications, the problem of manufacture has in the past sometimes discouraged the full exploitation of this simple concept. However, currently numerically-controlled machines make it possible to fabricate a development or 'breadboard' model reflector from solid material under control of a paper tape. Ultimately, the design may be produced in quantity using a mould or template which is itself manufactured in this way.

Recently, this approach has found application in the field of satellite communciations. For example, the MARECS L-band antenna uses a contoured reflector. The polar diagram is shaped to provide a degree of compensation for the increased length of propagation paths terminating near the edge of the earth's disc.

To tackle the synthesis problem for a shaped beam antenna, what is required first and foremost is an efficient analytical technique to predict gain and polar diagrams. It then becomes possible to devise algorithms whereby an 'initial guess' reflector profile is modified iteratively until the best possible approximation to the desired pattern is obtained. Some aspects of this problem will be addressed in later chapters of this book.

In certain applications, the polarisation properties of the reflector antenna are of special importance. For example, in the usual monopulse arrangement for a radar tracker, the antenna is required to generate a 'difference' pattern with a central tracking null. The corresponding crosspolar pattern usually tends to maximise on axis. Since a cross-polarised component of the radar echos is inevitable, this could constitute a major source of tracking error.

A recent satellite communication system concept is known as 'frequency re-use'. This enables the effective channel capacity of a satellite to be doubled by using two orthogonal polarisations for each frequency band. The necessary orthogonality can be between two perpendicular linear polarisations or left-hand and right-hand circular polarisation. In either case, one antenna is used to receive, transmit, and separate the two polarisations. Under these conditions, any crosspolar response that the antenna may have will tend to create 'crosstalk' between nominally separate channels (albeit that in the case of binary modulation techniques, such crosstalk will appear as noise).

The specification for a system of this kind typically involves a maximum permissible 'wanted to unwanted ratio' of the order of 30 dB. For the antenna on the satellite, this specification must be maintained 'off-axis' at points within the earth coverage zone, and implies crosspolar lobe levels of −35 dB or better. For the ground terminal antenna, polarisation purity plays a major role only in relation to the near axis response within the tracking cone. Nevertheless, the requirements are still stringent, and will inevitably be even more exacting for antennas which are to be used as standards to test the communication antennas. This applies both in relation to ground range tests and in-flight tests when the satellite antenna is tested as a part of the complete communication system.

The crosspolar response of a practical reflector antenna is an amalgamation of contributions due to a number of individual sources. These include: orthogonal mode coupling in the feeder waveguide, feed horn and orthomode transducer (if used), polariser response (for circularly polarised applications), feed horn crosspolar response, and crosspolarisation associated with induced currents in the reflector itself or in the feed support structure.

Analytical prediction of the crosspolar response involves in the first instance accurate modelling of the vector radiation pattern of the primary feed horn, preferably using measured data. For a front-fed antenna the horn in itself frequently accounts for a major portion of the overall crosspolar response. Its crosspolar pattern is nevertheless modified by the focusing action of the reflector. There is also a residual crosspolarised signal associated with the vector properties of the induced current on the reflecting surface. The two effects (reflector crosspolar efficiency and depolarisation, respectively) may be evaluated without recourse to any new analytical techniques special to the crosspolar problem. In fact crosspolar pattern information is obtained automatically as a by-product of a reflector scattering treatment of the vector diffraction type, as will be described in Chapter 6.

For circular polarisation, an additional mechanism which can degrade polarisation purity can sometimes arise via a second-order multiple scattering process. Thus for example, a LHCP field incident on a feed horn can be re-reflected at the LHCP port of the orthomode transducer and re-emerge as LHCP. This is in contrast to scattering from a reflector which introduces a polarisation inversion. Thus that portion of the scattered field of a reflector which re-impinges on the feed horn excites 'unwanted hand' CP to the extent that the OMT ports are imperfectly matched.

Usually, it will be necessary to make some modification to the reflector surface to suppress this interaction. The use of a central reflecting disc or vertex plate is well known. However, in some applications, it may be preferable to introduce a special distortion of the reflecting surface. In this way, the interaction can be suppressed with minimal interference to the focusing or beam-shaping properties of the reflector. An analytical approach to the feed horn interaction problem will be described in Chapter 9.

Other contributions to the overall crosspolar response can be more difficult to identify or treat analytically. At the present time no really satisfactory proven technique exists for estimating feed support structure crosspolarisation by analytical means, and only a few simple guidelines are available to suggest how this effect might be minimised.

Of course the most direct approach is to do away with blockage effects by using an offset feed. It has been demonstrated on a number of occasions that total crosspolar response can be predicted quite accurately from a theoretical model, when the reflector is offset-fed. Offset-fed reflectors introduce a form of 'geometrical optics depolarisation'. The properties of these reflectors with nominal linear and circular polarisation excitation will be discussed in a later part of this book.

The preceding brief review of reflector antenna applications has been included to illustrate the kind of features that need to be incorporated in an analytical approach. Specifically, provision needs to be made for:

(*a*) introduction of measured data (e.g. feed or feed-system patterns, measured profile surface co-ordinates) in analytical model (*b*) evaluation of successive scattering processes where a number of reflectors are used

(*c*) modelling of the three-dimensional features of antenna geometries

(*d*) treatment of 'wide-angle' scattering processes: this includes the calculation of defocused scattered patterns which are inherently wide angle, and the estimation of low-level wide-angle radiation from a focused antenna

(*e*) faithful modelling of the vector nature of electromagnetic fields, to allow the polarisation properties of the antenna to be evaluated

(*f*) synthesis (e.g. the determination of a feed/reflector combination that will realise a given radiation pattern) in addition to analysis.

The principal object of this book is to outline certain recent developments in the field of numerical methods applied to reflector antenna problems, pertinent to the requirements listed above. As such the monograph makes no pretence to be a comprehensive design handbook for the many applications of the reflector antenna. Likewise there will be no attempt to describe in any substantial detail all the specific variants of the reflector antenna prototype, together with their capabilities, advantages and disadvantages. Rather, the emphasis will be placed primarily on the underlying basis of the numerical methods, and their implementation. Of course, the methods are useful only to the extent that they are orientated towards real requirements of the antenna designer. Bearing this in mind, details of specific applications will be included whenever these are relevant to the direction taken in developing the theoretical analysis.

Implicit in many of the types of calculation to be described in this book is the scattering problem of an electromagnetic wave and a finite curved conducting surface. Usually it is not feasible to obtain an exact solution to an electromagnetic boundary value problem of this type. However, as is well known, various approximate solutions may be formulated. In particular, three types of treatment accurate to the order of vector diffraction* can be cited as follows:

(i) *PO/KHS Integral:* This is often referred to as the 'induced current' method. The currents on the reflecting surface are approximated using the physical optics (PO) principle, and the radiation due to these currents is then obtained by numerical evaluation of a standard radiation integral (The Kirchhoff–Huygens–Silver integral).

* The term 'diffraction' is sometimes loosely used, but in this book will be taken to imply those (usually frequency dependent) aspects of the solution of an electromagnetic problem not predicted by the ray propagation specified by optics

This gives a fairly complete description of wide-angle scattering, polarisation, and near-field effects. For some types of problem, however, requirements for computer time and storage can tend to become excessive.

(ii) *GTD:* The geometrical theory of diffraction is a more recent development, involving an extension of geometrical optics (GO) by the inclusion of additional diffraction rays. The method is particularly suited to problems with multiple reflection paths and obstacles having sharp edges.

The GTD approach avoids the numerical integrations inherent in the induced-current method, and hence can be very economical in respect of computer time/storage requirements, for simple reflector types such as the circularly symmetric paraboloid or hyperboloid. However, for other geometries, the computer algorithms required may be quite complex, and will be individual to each particular problem.

(iii) *PO/SWE:* This technique starts with a calculation of induced currents (physical optics) but matches a spherical wave modal expansion (the SWE) to these currents rather than evaluating a diffraction integral. As a result, a rigorous vector electromagnetic representation is obtained in a form suited to efficient machine computation. The method leads naturally to universal 'modular' algorithms, applicable to a wide variety of different antenna configurations. A special feature of the SWE is its adaptability to inverse and synthesis problems.

In the following chapter the alternative solutions of the fundamental scattering problem will be set down in more detail. The remainder of the book will largely be biased towards the approach favoured by the author, namely the modal-expansion method.

Chapter 2

Diffraction theories as applied to reflector antennas

2.1 Geometrical optics

In the ballistic theory of light the propagation of electromagnetic energy is explained in terms of the kinematics of photon particles. Whether or not this hypothesis is nearer to physical reality than that of wave propagation, it does lead to a particularly useful concept. This is the concept of an optical ray as the trajectory of a light particle. The tangent to a ray at any point represents equally the local direction of propagation, or the normal to the wavefront. Also associated with the ray at any point is its intensity. Proportional to the latter are the ray field vectors $\boldsymbol{E}$ and $\boldsymbol{H}$. These are also equal in magnitude,* and mutually orthogonal. The vector components of $\boldsymbol{E}$ are associated with an additional property of the ray, polarisation. This is often unimportant at optical frequencies but is usually of interest at microwave frequencies.

The fundamental laws specifying optical ray propagation where all radiating or scattering surfaces are very large in comparison to a wavelength are embodied formally in the 'eikonal' and 'transport'

* This book will be concerned exclusively with propagation of electromagnetic waves in free space or regions containing metallic scatters. It is therefore convenient to suppress the free-space wave impedance constant (377 Ω). The omission can be justified if also 'natural' units of E and H, equivalent to $(377)^{0.5}$ volts/meter and $(377)^{-0.5}$ amps/meter in the MKS case are defined. If the length unit is assigned to be one wavelength, we have the 'natural electromagnetic unit' or NEU. Thus E and H become quantities of the same order of magnitude when measured in NEU. Notwithstanding this, the product of E and H has most of the properties associated with conventional field units, being measured in watts per square wavelength. In the treatments of scattering problems described in this book, unit power has been taken as a starting point, e.g. 1 watt input to the feed horn when evaluating the pattern of a reflector antenna. Under these conditions the solution in NEU is independent of the unit of power adopted. That is, the NEU effectively becomes a dimensionless quantity in no way dependent on the MKS system.

equations.[1, 2] These concepts are also applicable at microwave frequencies with the proviso that diffraction effects are then more likely to be significant because of the longer wavelength. For antenna scattering problems ordinarily it is only necessary to treat free-space propagation, with straight-line ray paths, together with reflection at perfect conductors.

In this latter case the more formal generalisations of References 1 and 2 are unnecessary and the relevant laws can be stated most simply as follows: a ray incident on a reflecting surface will re-emerge as a reflected ray which is specifed by either

(*a*) *Snell's law:* angle of incidence = angle of reflection measured from surface normal; incident ray, normal, and reflected ray coplanar.

(*b*) *Fermat's principle:* if a ray propagates between two points A and B via a point C on a reflector, the path length ACB is stationary with respect to displacement of C.

In antenna applications, it is usually preferable to make use of Snell's law when the surface normal direction cosines are known explicitly. However, where they are known implicitly via specified numerical data for the profile co-ordinates, it is sometimes advantageous to use the alternative principle of Fermat.

The intensity of a ray is preserved at the point of reflection, although the tangential E vector undergoes a 180° phase change. Along a ray, the intensity varies in sympathy with the local ray flux or number of rays per unit area measured normally to the ray path. The phase also varies, in the manner of a free-space plane wave aligned locally with the ray.

In microwave antenna applications, geometrical optics (GO) techniques are most useful where energy is dispersed over a wide solid angle (e.g. scattering from a Cassegrainian subreflector). The GO solution is of course approximate but is usually relatively simple to apply. However, if there is a caustic present, e.g. region where rays run parallel over a long distance or converge to a point or a line, GO will not adequately describe the field behaviour. The focused parabolic reflector constitutes an example of this latter type of problem.

2.2 Aperture field and physical optics diffraction

2.2.1 Vector diffraction integrals

The radiating properties of any primary source (e.g. horn antenna) or secondary radiator (e.g. reflector) can be calculated directly from the electric currents which are physically responsible for initiating the radiation. Alternatively, they can be calculated from the fields present on a closed 'secondary wavefront' surface surrounding the antenna. There are various integral formulations applicable, all of which arise essentially from the application of Green's vector identities to Maxwell's differential equations.[3] They will be referred to in this book under the general description of Kirchhoff–Huygens–Silver (KHS) diffraction integrals.

Thus if the radiating currents are supported either on a closed or an open surface S_i, and the displacement of the observation point is $r\boldsymbol{a}_r$ relative to a general point on S_i, whereas relative to a fixed co-ordinate origin in the vicinity of the source the displacement is $R\boldsymbol{a}_R$, then (see Fig. 2.1).

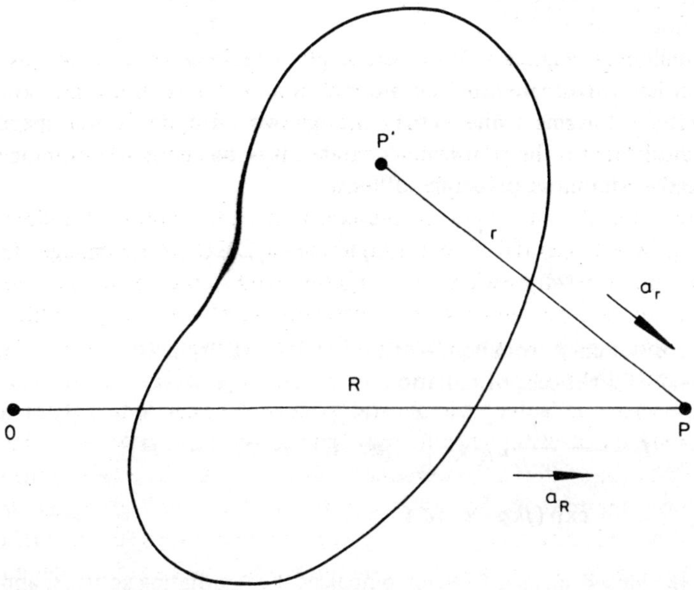

Fig. 2.1. *Radiating source and observation point*

$$\boldsymbol{E} = \frac{j}{4\pi k}\int_{S_i}\{(1+jkr+k^2r^2)\boldsymbol{i} - -(3+3jkr-k^2r^2)(\boldsymbol{i}\cdot\boldsymbol{a}_r)\boldsymbol{a}_r\}\frac{\exp(-jkr)}{r^3}\,dA \qquad (2.1)$$

(***E*** at arbitrary range)

$$H = -\frac{1}{4\pi}\int_{S_i} a_r \times i \left(\frac{1+jkr}{r^2}\right)\exp(-jkr)\,dA \tag{2.2}$$

(H at arbitrary range)

$$E = j\frac{k}{4\pi R}\exp(-jkR)\int_{S_i} a_R \times (a_R \times i)\exp(jk\boldsymbol{\rho}\cdot a_R)\,dA \tag{2.3}$$

(far field pattern, valid in the limit $R \to \infty$; $\boldsymbol{\rho}$ is a generating vector for the radiating surface S_i, measured from the fixed co-ordinate origin).

The vector i in these equations is the linear current density, e.g. current per unit of a length measured along the surface and normal to i. The integrands of both eqns. 1 and 2 exhibit poles corresponding to the condition source point → observation point. Terms up to $1/r^2$ are nevertheless integrable. However, eqn. 1 gives rise to numerical difficulties when evaluating fields very near to the source. This is due to the two $1/r^3$ terms which result in large, almost equal and opposite, contributions to the integral. In this case it is better to use the alternative and more fundamental form

$$E = \frac{1}{4\pi jk}\int_{S_i}\left\{k^2 i + \nabla\cdot i\left(\frac{1+jkr}{r}\right)a_r\right\}\frac{\exp(-jkr)}{r}\,dA \tag{2.4}$$

Eqn. 1 is obtained from eqn. 4 after integration by parts.

Finally for the case of radiation by 'secondary wavefront' fields E, H:

$$E = -\frac{jk}{4\pi R}a_R \times \int_{S_{sec}} \{n \times E - a_R \times (n \times H)\}\exp(jk\boldsymbol{\rho}\cdot a_R)\,dA \tag{2.5}$$

Here the closed surface $\int_{S_{sec}}$ encompasses the radiating sources, and has an outward unit normal vector of n.

2.2.2 *Signature of a current sheet source*

The presence of an electromagnetic source in a free space environment is announced by its characteristic field discontinuities or 'signature'. These discontinuities, impressed on an otherwise analytic solution of Maxwell's equations, may be thought of as initiating the radiated fields. In fact, it is sometimes advantageous to make use of this concept directly rather than attempt to apply the vector integrals of eqns. 1–5, as will become evident in later sections of this book.

For a current sheet source the relevant field discontinuities may be deduced directly from Maxwell's equations and are

$$D_{-+}\boldsymbol{H} = -\boldsymbol{n} \times \boldsymbol{i} \tag{2.6}$$

$$D_{-+}\boldsymbol{E} = \frac{j}{k}(\boldsymbol{\nabla} \cdot \boldsymbol{i})\boldsymbol{n} \tag{2.7}$$

and

$$D_{-+}\left[\frac{\partial \boldsymbol{E}}{\partial n}\right] = jk\boldsymbol{i} + \frac{j}{k}\,\boldsymbol{\nabla}(\boldsymbol{\nabla} \cdot \boldsymbol{i}) \tag{2.8}$$

Here the operator D_{-+} implies a difference of fields at immediately opposite points on either side of the source sheet. The sign convention is such that the field at the point for which $\boldsymbol{n}$ is an inward normal is to be subtracted. Eqs. 6—8 give the *only* discontinuities of the electromagnetic fields and their derivatives for an electric current source. The appropriate relationships for a magnetic current source $\boldsymbol{i}_M$ follow from the principle of duality, as described in the next Section.

2.2.3 *Electromagnetic duality*

Maxwell's equations possess an inherent symmetry with respect to E and H. Thus, corresponding to any particular solution there always exists a dual solution with the roles of these two vectors interchanged. Specifically the dual solution is obtained by substituting $\boldsymbol{H}$ in place of $\boldsymbol{E}$, $-\boldsymbol{E}$ in place of $\boldsymbol{H}$, and $\boldsymbol{i}_M$ in place of $\boldsymbol{i}$.

The use of the duality concept enables the vector diffraction integrals of eqns. 1–4 to be extended to the case of magnetic current sources $\boldsymbol{i}_M$. The latter usually arise in the context of the equivalence principle to be described in the next Section.

2.2.4 *Equivalence principle*

According to the electromagnetic equivalence principle[3] those sources interior to a specified 'secondary wavefront surface' S_{sec} can be replaced by sources on S_{sec} of

$$\left.\begin{aligned} \boldsymbol{i} &= \boldsymbol{n} \times \boldsymbol{H} \\ \boldsymbol{i}_M &= -\boldsymbol{n} \times \boldsymbol{E} \end{aligned}\right. \tag{2.9}$$

from the point of view of exterior radiation. Here $\boldsymbol{E}:\boldsymbol{H}$ constitutes the fields that would be launched by the interior sources at S_{sec}, and $\boldsymbol{n}$ is a unit normal vector directed towards the exterior region.

It is instructive to consider the special case where the interior source is a thin reflector and the closed surface S_{sec} is collapsed to take up the contour of this reflector on both the illuminated and shadow sides. The field $\boldsymbol{E}:\boldsymbol{H}$ is then just the scattered field $\boldsymbol{E}_s:\boldsymbol{H}_s$ due to the induced currents. $\boldsymbol{E}_s$ is continuous on either side of the reflector so that the shadow magnetic currents are equal and opposite to the illuminated side magnetic currents. However, $\boldsymbol{H}_s$ is discontinuous according to eqn. 6, resulting in a net equivalent electric current of

$$\boldsymbol{i}_{eq} = -\boldsymbol{n} \times \{\boldsymbol{n} \times (\boldsymbol{i}_{il} + \boldsymbol{i}_{sh})\} = \boldsymbol{i}_{il} + \boldsymbol{i}_{sh} \tag{2.10}$$

This current may be considered to be distributed over an open surface S_i constituting the illuminated side of the reflector. It must be

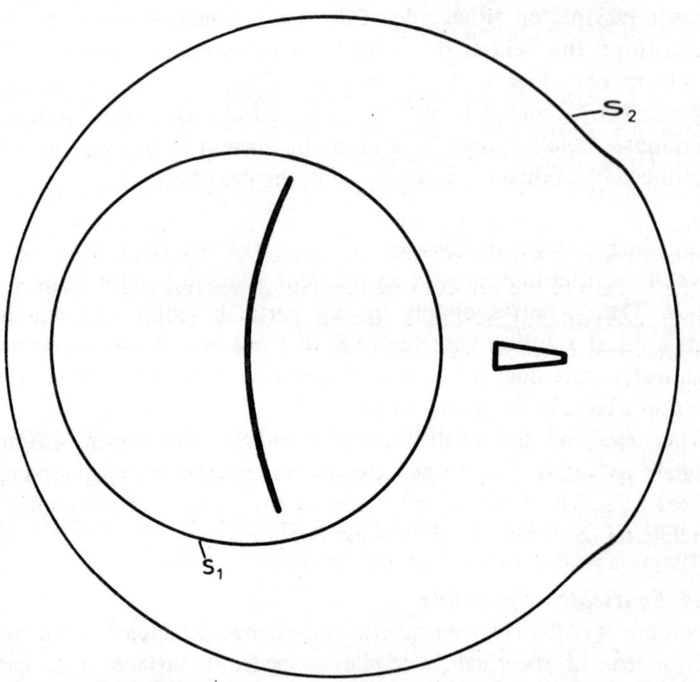

Fig. 2.2. *The equivalence principle applied to a reflector antenna*

remembered that $\boldsymbol{i}_{eq}$ does not launch fields which in themselves satisfy boundary conditions at the conducting surfaces of the reflector. Rather the equivalent current is to be associated with the scattered fields, whereas the total fields at the surface also include the incident fields.

The latter amount to the fields of the primary source horn for the usual front-fed reflector antenna.

The interpretation of equivalence in the particular context of a reflector antenna can be dependent on the meaning assigned to $\boldsymbol{E}:\boldsymbol{H}$ in eqn. 9. Thus if S_{sec} encloses *both* the reflector and the feedhorn (S_2 of Fig. 2.2), then the equivalent currents will radiate the total fields if $\boldsymbol{E}$ and $\boldsymbol{H}$ themselves are total fields. Alternatively, they will radiate the scattered field of the reflector if $\boldsymbol{E}:\boldsymbol{H}$ are scattered fields, and the primary feed fields if $\boldsymbol{E}:\boldsymbol{H}$ are primary feed fields. However, if for example S_{sec} encompasses the reflector only (S_1 of Fig. 2.2), then $\boldsymbol{E}:\boldsymbol{H}$ can be equally scattered fields or total fields, the corresponding equivalent currents launching only scattered fields in either case.

2.2.5 Radiation from a planar surface

2.2.5.1 Principle of images: The integration surface S_{sec} must in general be closed. However, one possible choice is the infinite plane $z = 0$. This divides free space into a region $z < 0$ containing the sources, and a region $z > 0$ in which the radiated fields are to be calculated.

Special properties pertaining to this planar secondary wavefront problem may be deduced from the principle of images. Thus, let $\boldsymbol{E}:\boldsymbol{H}$ be the fields of the sources when radiating into free space. Then assume a thin conducting screen is placed at $z = 0$ (Fig. 2.3). The screen introduces image sources which radiate fields in the rear, $z < 0$ half-space of

$$\boldsymbol{E}'(x,y,-z) = 2[\boldsymbol{a}_z \cdot \boldsymbol{E}(x,y,z)]\,\boldsymbol{a}_z - \boldsymbol{E}(x,y,z) \qquad (2.11)$$

$$\boldsymbol{H}'(x,y,-z) = \boldsymbol{H}(x,y,z) - 2[\boldsymbol{a}_z \cdot \boldsymbol{H}(x,y,z)]\,\boldsymbol{a}_z \qquad (2.12)$$

By way of contrast, for the forward halfspace the screen must launch scattered fields which just cancel the incident fields, e.g.

$$\boldsymbol{E}''(x,y,z) = -\boldsymbol{E}(x,y,z) \qquad (2.13)$$

$$\boldsymbol{H}''(x,y,z) = -\boldsymbol{H}(x,y,z) \qquad (2.14)$$

From eqn. 12, the total tangential magnetic field on the illuminated surface of the screen is just twice the incident field $\boldsymbol{H}$. From Maxwell's magnetomotive force law, the screen current is

$$\boldsymbol{i} = 2\boldsymbol{n}_1 \times \boldsymbol{H}(x,y,0)$$

This current must radiate the field $\boldsymbol{E}'':\boldsymbol{H}''$ for $z > 0$; consequently an electric current

$$\boldsymbol{i} = 2\boldsymbol{n}_2 \times \boldsymbol{H}(x,y,0) \qquad (2.15)$$

radiates the field $E:H$ and is equivalent to the rear halfspace sources for the purpose of calculating the forward halfspace fields.

Eqn. 15 represents an alternative to the equivalent sources of eqn. 9 for a plane aperture. In fact the radiated fields can be calculated from any linear combination of a total weight of unity for eqns. 15 and 9.

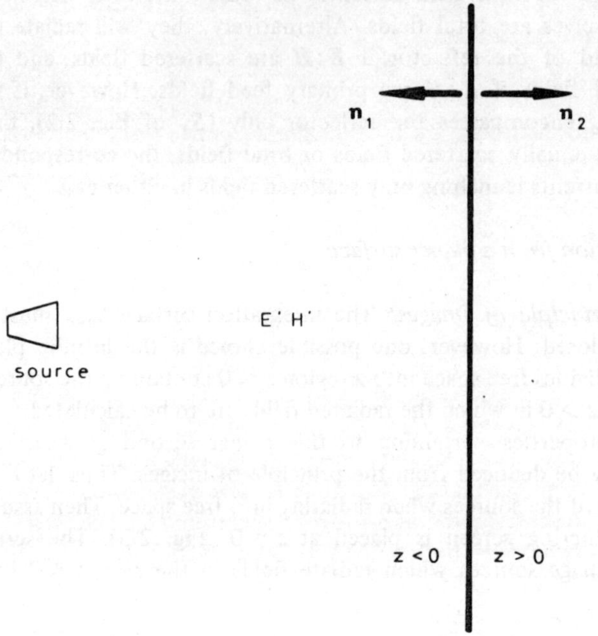

Fig. 2.3. *Scattering by a plane screen*

2.2.5.2 Plane wave spectrum: In the plane-wave spectrum technique of Booker and Clenmow,[4] the fields in the forward halfspace are calculated from a planar magnetic current equivalent source $-2\boldsymbol{n}_2 \times \boldsymbol{E}\ (x, y, 0)$. Thus from eqn. 5 it may be deduced that the far-field radiation pattern is

$$E(\Theta, \Phi) = jk \cos \Theta \frac{\exp(-jkR)}{R} A \tag{2.16}$$

where

$$A(k_x, k_y) = \frac{1}{2\pi} \int_{-\infty}^{\infty} \int_{-\infty}^{\infty} E_t(x, y, 0) \exp(-jk_x x - jk_y y) dx dy \tag{2.17}$$

and

$$k_x = k \sin \Theta \cos \Phi \tag{2.18}$$

$$k_y = k \sin \Theta \sin \Phi \tag{2.19}$$

The 't' suffix of eqn. 17 indicates that only the 'in plane' component of E contributes to the radiation, as implied by an $n_2 \times$ operation.

It is now apparent that for the planar aperture case the vector diffraction integral amounts to just a two-dimensional Fourier transform. Therefore an inverse transform must also exist, specifying $E_t(x,y,0)$. It may be shown that[4] this transform may be extended to the entire forward halfspace by introducing an $\exp(j\sqrt{\alpha}z)$ factor in the integrand, where

$$\alpha = k^2 - k_x^2 - k_y^2 \tag{2.20}$$

Thus

$$E_t(x,y,z) = \frac{1}{2\pi} \int_{-\infty}^{\infty} \int_{-\infty}^{\infty} A(k_x, k_y) \exp(jk_x x + jk_y y + j\sqrt{\alpha}z)dk_x dk_y \tag{2.21}$$

Values of k_x, k_y corresponding to visible space, e.g. real Θ, Φ in eqns. 18 and 19, give $\alpha > 0$ in eqn. 20. However, the integration limits of eqn. 21 imply that 'imaginary space' waves with $\alpha < 0$ must also be considered when calculating E_t. That is, the exponential factor of eqn. 21 may be interpreted as an angular spectrum of plane waves, and this spectrum includes both propagating waves and waves which are evanescent. The evanescent waves decay with z according to $\exp(j\sqrt{\alpha}z)$, with $j\sqrt{\alpha}z$ a negative real quantity.

Eqns. 16–21 assume that the $z = 0$ plane lies in free space. However, in one application of interest, a conducting screen containing a cut-out is placed at $z = 0$. The equations remain valid provided that $E_t(x,y,0)$ is interpreted as the *total* of the field incident on the screen and its scattered field. Since this total field is zero on the conducting surface, the integration limits in eqn. 17 now need include only the cut-out or 'aperture'. This type of formulation is particularly useful for treating horn antennas having an aperture flange. Although the latter is usually only about $\lambda/4$ wide, the infinite screen model nevertheless is a good approximation.

2.2.6 Principles of physical optics and local plane-wave reflection

One aspect of a scattering problem is to determine either the induced currents present on a reflector (induced current method) or a set of equivalent radiating sources positioned just above its surface (aperture field method). In either case a rigorous solution for a specified incident field $E_{inc} : H_{inc}$ will in general only be possible via an integral equation (see Section 2.4).

The physical optics (PO) and local plane-wave-reflection (LPWR) approximations follow directly from the imaging properties of a plane screen (Section 2.2.5.1). Thus for this problem

$$i = 2\boldsymbol{n} \times \boldsymbol{H}_{inc} \tag{2.22}$$

represents the induced current on the illuminated side of the screen, where $\boldsymbol{n}$ is drawn towards the source of illumination. The 'shadow side' induced current is zero for an infinite screen; eqn. 10 then demonstrates that the illuminated side induced currents and the equivalent current sheet used as a basis for vector diffraction integral calculations are one and the same for this case.

The scattered field just above the surface of the screen is

illuminated surface

$$\boldsymbol{E}_s' = 2(\boldsymbol{n} \cdot \boldsymbol{E}_{inc})\boldsymbol{n} - \boldsymbol{E}_{inc} \tag{2.23}$$

$$\boldsymbol{H}_s' = \boldsymbol{H}_{inc} - 2(\boldsymbol{n} \cdot \boldsymbol{H}_{inc})\boldsymbol{n} \tag{2.24}$$

shadow surface

$$\boldsymbol{E}_s'' = -\boldsymbol{E}_{inc} \tag{2.25}$$

$$\boldsymbol{H}_s'' = -\boldsymbol{H}_{inc} \tag{2.26}$$

Eqns. 22–26 are exact for a wave of arbitrary field distribution incident on an infinite plane conductor. However, even in the context of this canonical problem, it is necessary to assume that source – scatterer interaction is negligible. Eqn. 22 and eqns. 23–26 are commonly applied to reflector problems where the conducting surface is finite and curved; they then become the PO and LPWR approximations, respectively.

In the latter case, it is instructive to imagine the scattering process to take place in the following stages:

(*a*) primary field incident at reflector surface
(*b*) generation of induced currents
(*c*) secondary wavefront launched at reflector surface
(*d*) generation of scattered field distribution.

With the PO principle, the approximation enters in proceeding from steps (*a*)–(*b*). However, when LPWR is employed, the analysis progresses from (*a*)–(*c*) directly. Thus, although both principles are derived from the same canonical geometry, the induced current of eqn. 22 does not excite precisely the secondary wavefront of eqns. 23–26. Fig. 2.4 illustrates the distinction for the elementary example of a plane circular disc conductor excited by a normally incident plane wave. The tangential E-field is invariant in the region just above the disc according to eqns. 23 and 25, whereas a vector diffraction calculation starting from the induced currents of eqn. 22 evidences the diffraction ripples characteristic of an edge.

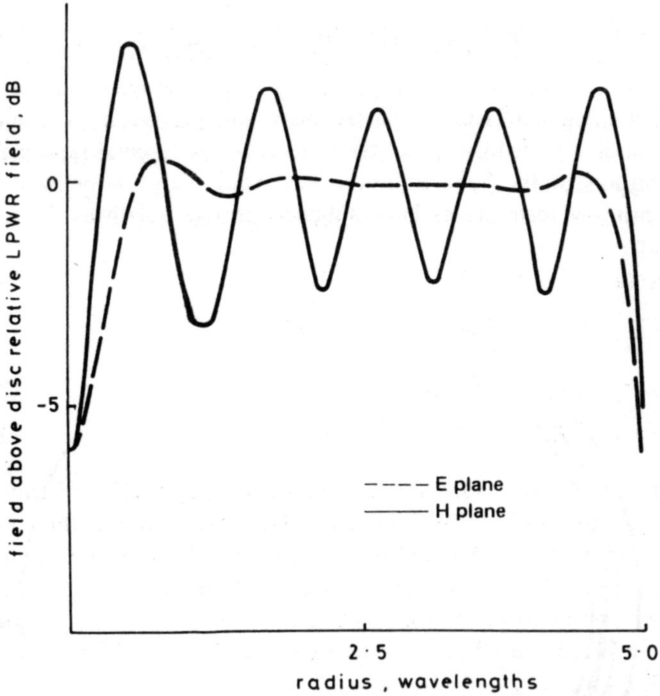

Fig. 2.4. *Illustrating the distinction between PO and LPWR: tangetial E field immediately above a 10λ disc excited according to PO by a plane wave*

2.2.7 *Superposition theorems*

In any scattering problem, the scattered field in itself does not constitute a complete solution. Thus, in the case of a PO treatment for a front-fed reflector antenna, the scattered field (as obtained by eqn. 22 in conjunction with eqn. 2 or 3) must be augmented by the free space radiated fields of the feed horn to obtain the total fields.

For a LPWR treatment, however, this simple type of superposition does not apply. Rather, with reference to Fig. 2.2, the total field is the sum of integrals over S_1 and S_2. These integration surfaces may be 'collapsed' to the configuration illustrated in Fig. 5. Now, because of eqns. 25 and 26, the integration over S_4 cancels with that over S_5 so that eqn. 5 becomes

$$E = \frac{-jk}{4\pi R} a_R \times \left[\int_{S_3} \{n \times E'_s - a_R \times (n \times H'_s)\} \exp(jk\rho \cdot a_R) dA \right.$$

$$\left. + \int_{S_6} \{n_f \times E_f - a_R \times (n_f \times H_f)\} \exp(jk\rho_f \cdot a_R) dA \right] \quad (2.27)$$

The first integral relates to the secondary wavefront $E'_s : H'_s$ emerging from the illuminated surface of the reflector. The second represents the contribution of the 'feed spill-over wavefront', $E_f : H_f$ being the free space primary feed fields. This spill-over term is relatively difficult to evaluate.

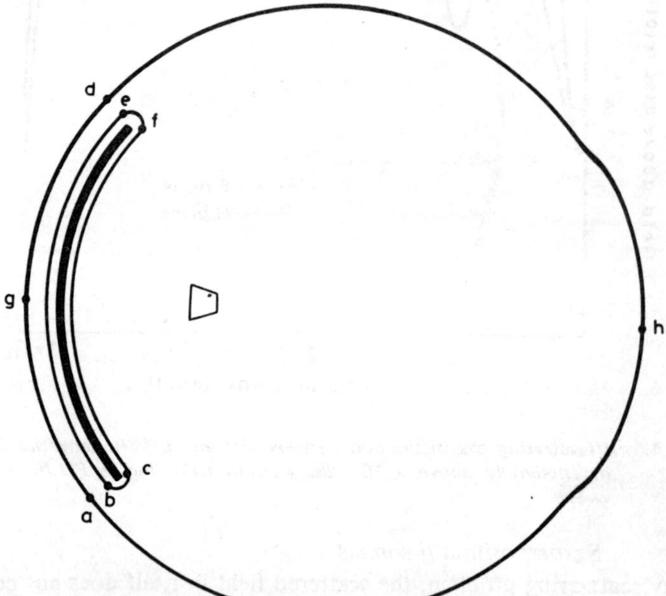

Fig. 2.5. *Superposition theorem for a reflector antenna*
S_3 = cf S_4 = be S_5 = agd S_6 = dha

For this reason, the LPWR method is fundamentally unsuitable for applications where the spill-over integral cannot safely be neglected, e.g. defocused scatterers and/or field points at large off-axis angles.

Frequently the LPWR method is used under conditions where S_3 is taken as a plane surface (the 'aperture plane') rather than the illuminated surface of the reflector. The field $\boldsymbol{E}_s : \boldsymbol{H}_s$ at S_3 is then calculated by geometrical optics principles from $\boldsymbol{E}_s' : \boldsymbol{H}_s'$ at the reflector surface. For example, for a focused parabolic reflector, the reflector surface field can simply be 'projected' to the aperture plane, on the basis of a parallel tube of rays emanating from the reflector (Fig. 2.6). However, the 'projection' of fields in this way involves further approximation in relation to the wide angle pattern.

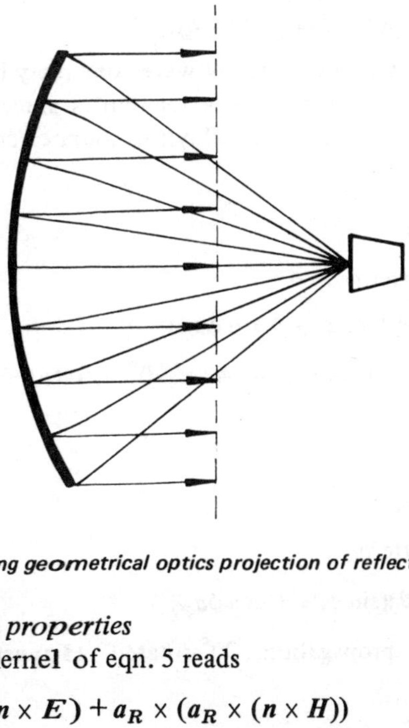

Fig. 2.6. *Illustrating geometrical optics projection of reflector near fields*

2.2.8 Polarisation properties

The polarisation kernel of eqn. 5 reads

$$-\boldsymbol{a}_R \times (\boldsymbol{n} \times \boldsymbol{E}) + \boldsymbol{a}_R \times (\boldsymbol{a}_R \times (\boldsymbol{n} \times \boldsymbol{H})) \qquad (2.28)$$

A local Cartesian co-ordinate system u, v, w may be defined such that the w-axis aligns with $\boldsymbol{n}$; the associated local spherical polar co-ordinate system is defined to be $\boldsymbol{r}, \theta, \phi$ where $\theta = 0$ aligns with $\boldsymbol{n}$. Now assume

$$\boldsymbol{E} = A\boldsymbol{a}_u$$

$$\boldsymbol{H} = B\boldsymbol{a}_v$$

Then, after substituting the identities

$$\boldsymbol{a}_u = \cos\theta \cos\phi \boldsymbol{a}_\theta - \sin\phi\, \boldsymbol{a}_\phi + \sin\theta \cos\phi \boldsymbol{a}_r$$

$$\boldsymbol{a}_v = \cos\theta \sin\phi \boldsymbol{a}_\theta + \cos\phi \boldsymbol{a}_\phi + \sin\theta \sin\phi \boldsymbol{a}_r$$

the two terms of eqn. 28 yield, respectively,

$$A(\cos\phi \boldsymbol{a}_\theta - \cos\theta \sin\phi \boldsymbol{a}_\phi) \tag{2.29}$$

and

$$B(\cos\theta \cos\phi \boldsymbol{a}_\theta - \sin\phi \boldsymbol{a}_\phi) \tag{2.30}$$

The quantities in the brackets constitute magnetic and electric dipole radiation patterns, respectively. If $A = B$ (e.g. $\boldsymbol{E} : \boldsymbol{H}$ is an element of wavefront behaving as a 'Huygens source' in relation to the chosen co-ordinate system) the overall elemental polarisation pattern takes on the form

$$(1 + \cos\theta)(\cos\phi \boldsymbol{a}_\theta - \sin\phi \boldsymbol{a}_\phi) \tag{2.31}$$

The radiation from any arbitrary wavefront may be treated in terms of Huygens sources. In general, a set of four is necessary at each field point. Thus 'reversed propagation' Huygens sources correspond to

$$\boldsymbol{E} = \boldsymbol{a}_u$$

$$\boldsymbol{H} = -\boldsymbol{a}_v$$

and radiate the pattern

$$(1 - \cos\theta)(\cos\phi \boldsymbol{a}_\theta + \sin\phi \boldsymbol{a}_\phi)$$

with a null on the forward axis. Also '90° – rotated' Huygens sources correspond to

$$\boldsymbol{E} = \boldsymbol{a}_v$$

$$\boldsymbol{H} = -\boldsymbol{a}_u$$

and radiate the pattern

$$(1 + \cos\theta)(\sin\phi \boldsymbol{a}_\theta + \cos\phi \boldsymbol{a}_\phi)$$

Finally, 'reversed propagation, 90°-rotated' Huygens sources correspond to

$$\boldsymbol{E} = \boldsymbol{a}_v$$

$$\boldsymbol{H} = \boldsymbol{a}_u$$

and radiate the pattern

$$(1 - \cos\theta)(\sin\phi \boldsymbol{a}_\theta - \cos\phi \boldsymbol{a}_\phi)$$

2.2.9 Physical optics and edge diffraction

When the fields of an antenna are calculated by vector diffraction methods, the solution may be thought of as constituting the GO fields (infinite frequency limit) plus a frequency-dependent perturbation. For example, in the case of a Cassegrainian subreflector, there are 'diffraction ripples' superimposed on the reflected GO ray cone and the slope of the pattern at the ray cone shadow boundary is finite (e.g. see Chapter 7). These diffraction perturbations are observed to be roughly proportional to the illumination level at the rim of the reflector.

In fact, an explicit dependence on fields at edges is immediately evident from the point of view of the geometrical theory of diffraction (Section 2.3). By way of contrast, at the stage of a canonical problem (Section 2.2.5), the PO does not admit the presence of any edge. Nevertheless, the edge is introduced at the stage of step (*d*) of Section 2.2.5. An asymptotic reduction of the PO vector diffraction integral[5] gives direct evidence of an edge illumination contribution. Moreover, the full PO solution encompasses diffraction mechanisms not accounted for by the standard GTD diffraction coefficients. An example is the so-called 'slope wave diffraction[6]' corresponding to the case where the primary feed horn pattern exhibits a null at the angle corresponding to the reflector rim.

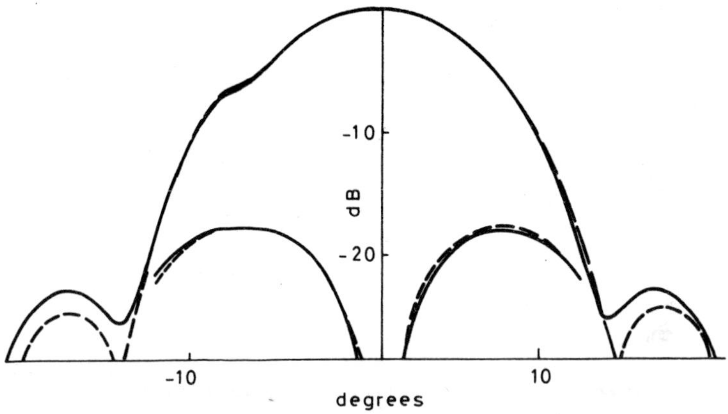

Fig. 2.7. *Copolar and crosspolar patterns for a 6λ offset fed reflector*
——— measured
- - - - predicted (PO)
H plane, symmetry plane

2.2.10 Limitations of physical optics

It is to be expected that the PO/induced current method will become less accurate for reflectors whose dimensions are not large in wavelengths (the influence of the edge is then increased). Equally, the accuracy might be expected to degrade when the local radius of curvature for the reflector surface is not everywhere large in comparison to a wavelength.

However, these considerations are seldom of concern in real practical applications. Thus for example Fig. 2.7 illustrates PO and measured patterns for an offset-fed reflector. The accuracy of the analytical prediction is compatible with the accuracy to which the primary feed horn pattern (used as input to the calculation) was specified, even though the diameter of the reflector was only 6λ. The discrepancy between theory and experiment is frequently more marked in the case of a centre-fed antenna. However, this is usually because of unaccountable blockage and multiple reflection effects, rather than any failure of PO.

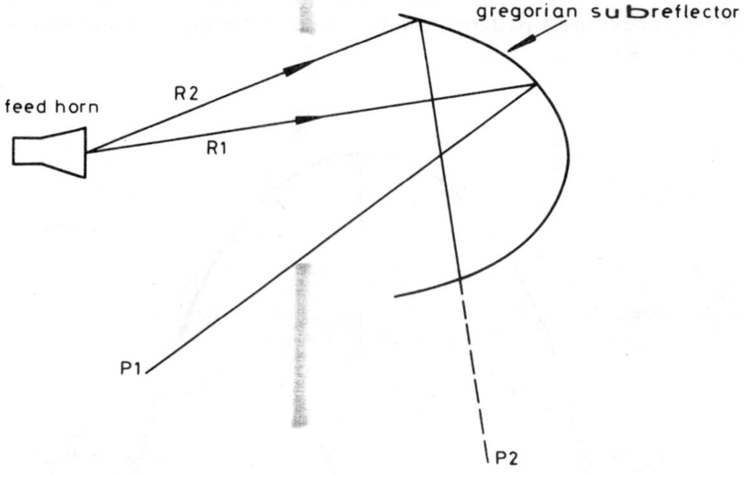

Fig. 2.8. *Reflector scattering problem in which self-shadowing occurs*

In Chapter 9 excellent agreement is demonstrated between the computed and measured V.S.W.R. performance of a Cassegrainian subreflector having matching rings. The latter constitute perturbations to the profile having radii of curvature of only λ/3. The analytical prediction of V.S.W.R. depends upon the calculation of a near-field scattered pattern of the subreflector using the PO/induced current method.

The chief limitation of PO relates to its adherence to the principle of superposition of scatterers. Thus, the PO scattering from two surfaces S_A and S_B will be the sum of the PO scattered fields with each surface present individually. As a consequence, PO cannot allow for multiple scattering interactions. For example, for the Gregorian subreflector of Fig. 2.8, one portion of the surface re-radiates onto another, as can be seen by tracing the GO rays. The PO/induced current method is often unsatisfactory when the reflector is self-shadowing.

2.2.11 Rotationally symmetric reflectors

In general, substituting the PO current in a vector diffraction formula such as eqn. 2 leads to an integral which is not amenable to purely analytical reduction and must be evaluated by numerical methods.[7] However, some useful simplification is possible in the important special case of a rotationally symmetric reflector and 'pseudo circularly-symmetric' feed system. The latter description implies that the feed pattern is of the general form

$$\boldsymbol{H}_f = f_1(\theta_f)\sin\phi_f \boldsymbol{a}_{\theta_f} + f_2(\theta_f)\cos\phi_f \boldsymbol{a}_{\phi_f} + f_3(\theta_f)\sin\phi_f \boldsymbol{a}_{r_f} \tag{2.32}$$

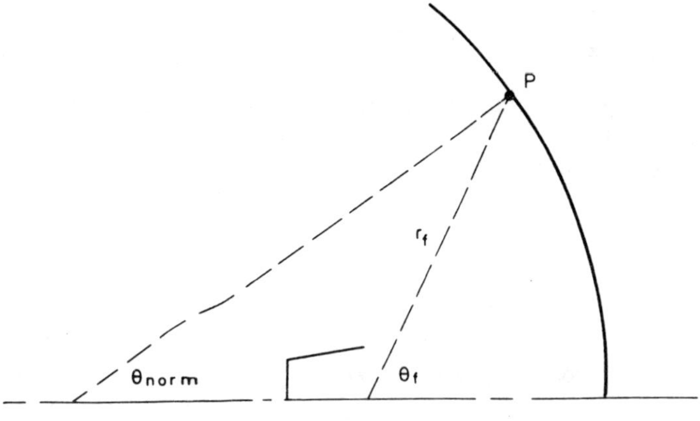

Fig. 2.9. *Geometry for application of KHS diffraction integral*

where r_f, θ_f, ϕ_f constitutes the feed spherical polar co-ordinate system of Fig. 2.9, $\theta_f = 0$ coinciding with the axis of revolution of the reflector. Once excited by the TE_{11} or TM_{11} circular waveguide modes, this sine/cosine type of ϕ dependence is preserved through any combination of coaxial, rotationally symmetric scattering boundaries.* Both

* A proof of this statement follows directly from the theory of spherical waves, to be introduced in the next chapter

the smooth wall and corrugated wall conical feed horns are examples.

Assuming $f_3 = 0$ in eqn. 32, e.g. no radial field component as for a far field pattern, and

$$\left.\begin{aligned} D/R &\ll 1 \\ 1/\mathrm{KR} &\ll 1 \\ d/R &\ll 1 \end{aligned}\right\} \quad \text{and} \tag{2.33}$$

where D is the diameter of the reflector and d its axial depth, then after some algebraic manipulation eqn. 2 may be approximated as

$$H \approx \frac{-jk \exp(-jkR)}{4\pi R} \int (a_R \times i) \exp\left(jk\left[\boldsymbol{\rho} \cdot a_R - \frac{x^2 + y^2}{2R}\right]\right) dA \tag{2.34}$$

Here $\boldsymbol{\rho}$ is a reflector surface generating vector measured from $x = y = z = 0$, and

$$x = \boldsymbol{\rho} \cdot a_x$$

$$y = \boldsymbol{\rho} \cdot a_y$$

the z-axis being the antenna boresight axis for the radiation pattern to be calculated.

By using standard Fourier-Bessel identities,[8] the 'ϕ' integration implicit in eqn. 34 can be eliminated,[9] leaving only the one-dimensional integrals in θ

$$I_1 = -\pi \int_0^{\theta_f} \{(J_0 + J_2) f_1(\theta_f) \cos\psi + (J_0 - J_2) f_2(\theta_f) \cos\theta_{norm}\} \mu \cos\Theta d\theta_f \tag{2.35}$$

$$I_2 = -2\pi j \int_0^{\theta_f} J_1 f_2(\theta_f) \sin\theta_{norm}\, \mu \sin\Theta d\theta_f \tag{2.36}$$

$$I_3 = \pi \int_0^{\theta_f} \{(J_0 - J_2) f_1(\theta_f) \cos\psi + (J_0 + J_2) f_2(\theta_f) \cos\theta_{norm}\} \mu d\theta_f \tag{2.37}$$

Here the argument of the Bessel functions J_0, J_1, J_2 is

$$u = kr \sin\theta_f \sin\Theta \tag{2.38}$$

and

$$\psi = \theta_f - \theta_{norm}, \tag{2.39}$$

$$\mu = r^2 \sin\theta_f \sec\psi \exp\left\{-jkr\left(\cos\theta_f \cos\Theta - \frac{r\sin^2\theta_f}{2R}\right)\right\} \tag{2.40}$$

The integrals of eqns. 35–37 can be evaluated by using a numerical quadrature rule such as Simpson's rule, in one dimension only (e.g. integration with respect to θ_f). The scattered pattern at the near field point R, Θ, Φ is then given by

$$E_s = -\frac{jk}{2\pi R}\{(I_1 + I_2)\cos\Phi \boldsymbol{a}_\Theta + I_3 \sin\Phi \boldsymbol{a}_\Phi\} \tag{2.41}$$

2.3 Asymptotic diffraction theories

2.3.1 Asymptotic physical optics

Excluding the special case of a focused region of the radiation pattern, the integrand of a vector diffraction integral such as eqn. 2 is always a highly oscillatory function of the integration variable. This property results from its phase term, which takes account of both the variation in path length from different points P on the reflector surface to the observation point, and also the phase variation of the reflector induced currents.

Whilst an oscillatory behaviour tends to hinder the application of a numerical integration rule, it permits the integral to be reduced analytically using the method of stationary phase.[10] The principles of GO may be derived in this way, as the high-frequency limit of vector diffraction theory. Moreover, extensions of the stationary phase method allow further, perturbation terms of the asymptotic solution to be identified. These may be associated with 'edge diffraction ray' contributions.[5, 11, 12]

For finite frequency and antenna dimensions, the stationary phase principle is an approximation, and consequently the APO technique is only valid for the calculation of a radiation pattern in restricted angular regions; in this respect it very closely parallels the geometrical theory of diffraction,[10] which will now be introduced.

2.3.2 Keller and uniform GTD

The GTD arises from a completely independent line of approach. It is postulated from the outset that the short-comings of GO can be remedied by introducing additional rays. It is assumed that these rays obey the propagation laws of GO rays such as Fermats principle (Section 2.1). The most important type of additional ray is the edge-diffraction ray. A single incident ray gives rise to a cone of such rays at a knife-edge (see Fig. 2.10).

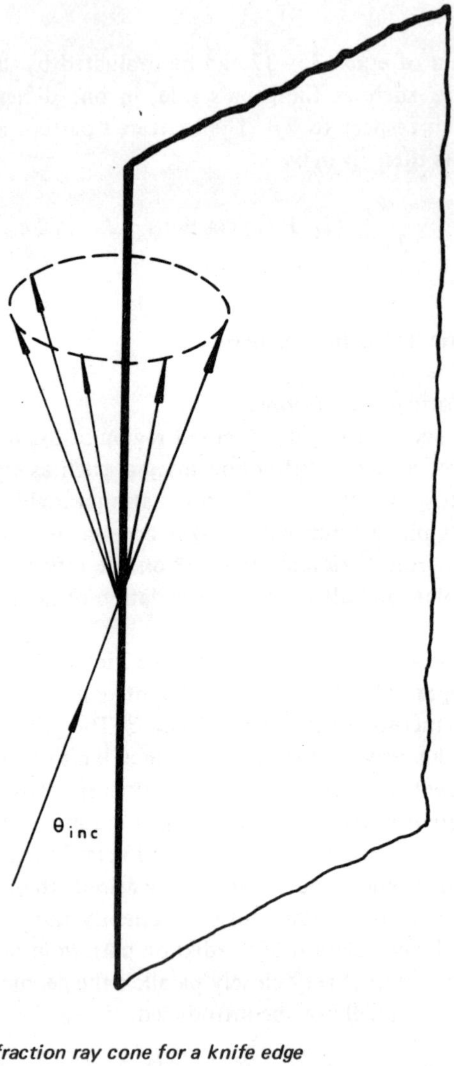

Fig. 2.10. *Diffraction ray cone for a knife edge*

Both qualitatively and quantitatively, the behaviour of a GTD ray is deduced from analysis of the appropriate rudimentary or 'canonical' problem. For 'knife-edge rays', the canonical problem comprises a plane wave incident on a conducting halfplane, a problem first solved by Sommerfeld.[13] His solution was phrased in terms of Fresnel integrals, but for the basic Keller GTD[14] these are replaced by their simpler 'far-field' asymptotic representations for the purposes of deriving 'diffraction coefficients'.

The latter, together with a geometrical factor, define the ratio of the amplitudes of the diffracted and incident rays. They thus constitute the principal vehicle by which the solution to the canonical problem is translated to the real application. The form of solution is sensitive to the polarisation of the incident wave. Thus, a 'soft' diffraction coefficient ('*s*' suffix in formulation to follow) corresponds to an incident E vector lying in the plane containing the incident propagation path and a perfect electric conductor knife-edge. A 'hard' diffraction coefficient ('*h*' suffix) is associated with an orthogonally polarised incident field.

These diffraction coefficients may be specified as[14]

$$D_{s,h}(\theta_{inc}, \phi_{inc}, \phi_{diff}) = \frac{1}{4\pi} \exp(j3\pi/4)\, \lambda^{1/2} \operatorname{cosec} \theta_{inc} \left\{ \sec\left(\frac{\phi_{diff} - \phi_{inc}}{2}\right) \mp \sec\left(\frac{\phi_{diff} + \phi_{inc}}{2}\right) \right\} \tag{2.42}$$

Here ϕ_{inc} and ϕ_{diff} correspond to the incident and diffracted rays of Fig. 2.10 in relation to a 'local' spherical polar co-ordinate system r, θ, ϕ. The latter has its z-axis aligned with the knife-edge and $r = 0$ at the local point of diffraction under consideration. The halfplane itself is defined by $\phi_{inc} = \phi_{diff} = 0$. θ_{inc} is the angle between the incident ray and the edge, and also the cone semiangle for the diffracted rays. The $\lambda^{1/2}$ factor in the formula is characteristic of edge diffraction, and ensures that the diffraction coefficient vanishes in the optical limit of zero wavelength. The upper sign in the {} brackets of eqn. 42 is to be taken for the case of the soft diffraction coefficient, and the lower sign for a hard diffraction coefficient.

The Keller formulation as outlined above would permit only a part of the field distribution to be calculated for a real scattering problem. For example, when applying the method to the reflector antenna of Fig. 2.11, it fails:

(*a*) in or near focused or caustic regions, (i) and (ii) of Fig. 2.11.
(*b*) at or near the incident field GO illumination/shadow boundary, e.g. $\phi_{diff} - \phi_{inc} \approx \pi$ for halfplane problem, region (iii) for Fig. 2.11.

(*c*) at or near the scattered field GO illumination/shadow boundary, e.g., $\phi_{diff} + \phi_{inc} \approx \pi$ for halfplane problem, region (iv) for Fig. 2.11.

Various attempts have been made to 'reform' the method to correct these deficiencies. Kouyoumjian's uniform diffraction[15] coefficients remedy (*b*) and (*c*). The total (GO + edge diffraction) fields are continuous across shadow boundaries with Kouyoumjian's theory. This property is achieved at the cost of some algebraic complexity and the necessity to use different formulations for different regions of the scattered pattern. An application to scattering from surfaces having rotational symmetry is given in Reference 16.

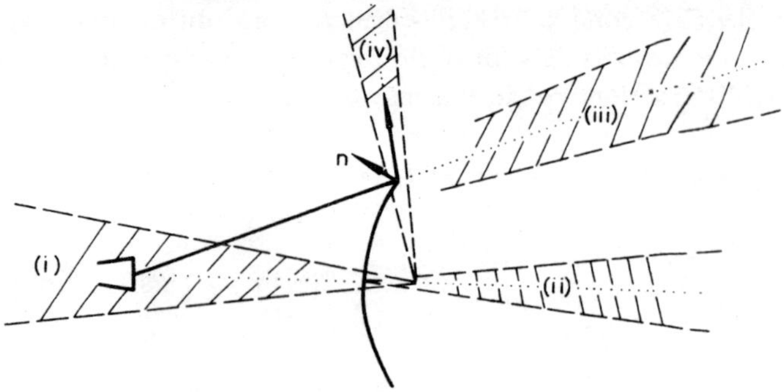

Fig. 2.11. ***Regions for which Keller GTD is invalid***

2.3.3 Comparison of asymptotic and physical optics theories

An advantage of GTD is that in some simple problems the numerical integrations inherent in the PO method may be avoided with consequent economies in computer CPU time. Furthermore GTD can take into account multiple scattering mechanisms if the appropriate additional rays are introduced. Fig. 2.12 illustrates two such problems,[17, 18] for which the GTD is probably the only feasible approach.

For some antenna geometries the ray tracing required for GTD may pose some problems, both in terms of formulating a computer algorithm and in terms of the computing requirements to carry out the calculation. In this context a new algorithm is ordinarily needed for each new geometry with the GTD, whereas a PO formulation tends to be more universal. Even when the uniform theory diffraction coefficients are used, the GTD solution is incomplete in the sense that the radiation pattern cannot be predicted in focused or caustic regions. The basic GTD is also not valid for the calculation of near-field patterns.[19]

Both GTD and PO are essentially high-frequency approximations, and improve in accuracy as the size of the scatterer in relation to the wavelength is increased. For PO, the approximation is confined to one stage of the analysis (the derivation of induced currents from incident fields). Its effect is to an extent 'traceable', either by estimating the 'error current' using an integral equation solution (Section 2.4.1) or by calculating the fields at the surface of the reflector using the uniform spherical wave expansion (Section 3.4).

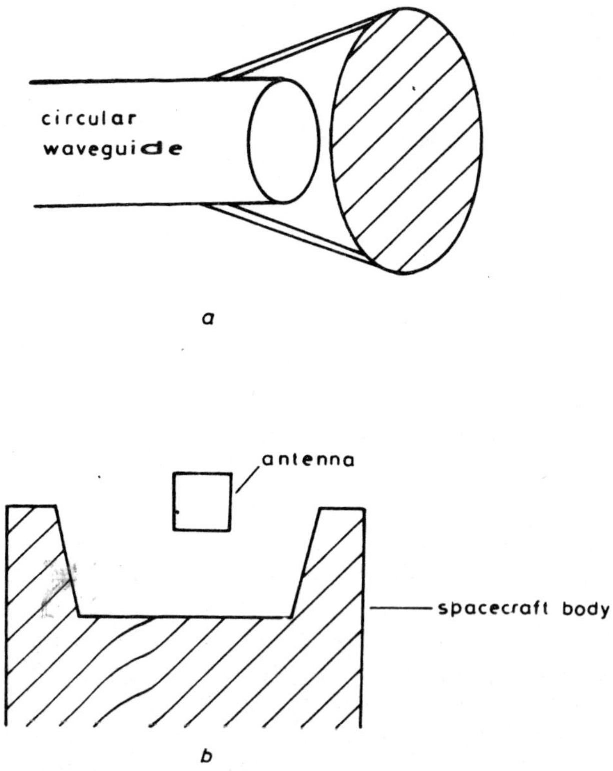

Fig. 2.12. *Two problems amenable to a GTD treatment*
a splash plate feed
b uhf antenna on spacecraft

For the GTD, the sources of error involved are less well defined. The GTD is certainly more rigorous than PO when applied to the original two- dimensional canonical problem of a halfplane. In the context of a reflector antenna application, however, approximation will enter in

(*a*) the formulation of the edge diffraction coefficients
(*b*) the use of GO ray tracing and propagation laws for diffraction rays
(*c*) the assumption of 'local excitation' for diffracted rays
(*d*) possible additional diffraction mechanisms not present in the canonical problem.

Recently it has been demonstrated[20] that in one particular antenna application (prediction of wide angle sidelobes of a parabolic reflector) the GTD is nevertheless more accurate than APO. A corollary would be that the PO interpretation of the knife-edge mechanism of diffraction is slightly less accurate than that of GTD.

The PO integral as already stated incorporates other mechanisms of diffraction. A comparison of overall accuracy with PO and GTD is possible in cases in which an exact integral equation solution is available. Where this has been done,[21, 22] no clear cut advantage has emerged for either method. In any event, in a real application other factors tend to be more important, such as the extent to which the modelling of the blockage of the feed support structure is realistic, and the accuracy of the feed system pattern data.

2.4 Exact solutions

2.4.1 The integral equation method

The 'integral equation' for a vector diffraction problem is simply a mathematical embodiment of the statement that the induced currents must launch fields which, together with the incident field, satisfy continuity relationships at all boundaries.

In general, either eqn. 4 or eqn. 2 may be used to predict fields at the surface of a conducting scatterer, forming the basis of an EFIE (electric field integral equation) or an MFIE (magnetic field integral equation), respectively. For a thin conducting sheet (e.g. a reflector), to use the MFIE it is necessary to distinguish between the magnetic field on either side of the sheet. By way of contrast, the basic EFIE

$$(n \times E_{inc})_P = \frac{1}{4\pi jk} \times \int \left\{ k^2 i_Q + \nabla_t \cdot i_Q \, \frac{1 + jkr_{PQ}}{r_{PQ}} a_{r_{PQ}} \right\} \frac{\exp(-jkr_{PQ})}{r_{PQ}} \, dA \tag{2.43}$$

works in terms of tangential electric field which is continuous through the sheet, and an open surface S_i containing the equivalent currents.

The bar on the integral sign indicates that the solution to be taken is mathematically the principal value[23] in respect of the pole of the integrand for source point $Q \to$ observation point P. The symbol ∇_t specifies a transverse (in relation to S_i) divergence operator.

On introducing a numerical integration rule, the RHS reduces to a weighted summation of i_Q values at grid points, for any specific P. Thus, by allowing P to coincide with each grid point in turn, a set of simultaneous equations could be obtained. In principle, these would suffice to solve uniquely for the grid point i_Qs. However, additional constraints have to be specified when the scatterer has sharp edges. Further difficulties arise when the geometry allows a closed cavity resonance at a frequency near to that for the calculation.[24]

In practice, a 'point matching' solution as outlined above is rarely used. Because of the implied presence of spatial derivatives of $\boldsymbol{i}$ via ∇_t, and the need to treat the pole of the integrand, it is better at the outset to introduce a specific analytic expansion for $\boldsymbol{i}$. This approach is known as the 'method of moments', the individual terms of the expansion being 'basis functions'. The method is well documented in the literature.[25, 26]

The basic EFIE of eqn. 43 neglects source-reflector multiple scattering. That is, $\boldsymbol{E}_{inc}$ should strictly be the field of the primary source allowing for any perturbation to the radiating properties of the latter due to the presence of the reflector. An extension to the method to treat the interaction problem is described in Reference 27.

The size of the scattering structure that can be treated by the MOM is frequently severely limited by difficulties in inverting large matrices with a digital computer. For three-dimensional scattering structures, this can mean that a scatterer of linear dimensions greater than about 2λ cannot be accommodated. Unfortunately, many of the problems that would most benefit from an MOM solution (e.g. a reflector *and* its feed and support structure) come into this category.

2.4.2 Iterative methods

The numerical evaluation of the surface fields at S_i for a given $\boldsymbol{i}$ can be a much more tractable task in computational terms than inverting a matrix to find the current distribution that will generate specified fields at S_i. This suggests building up a solution for $\boldsymbol{i}$ by a series of "hill-climbing' iterations, aiming to synthesise a surface electric field of $-\boldsymbol{E}_{inc}$. Until recently, however, attempts to implement this kind of technique have proved unsuccessful, probably because the exact $\boldsymbol{i}$ is nonanalytic at an H-plane edge.[13]

Mittra[28] has suggested a new technique which works in terms of the vector Fourier transform of $\boldsymbol{i}$. It remains to be demonstrated whether the method can be extended to configurations of interest in antenna engineering.

References

1. FELSEN, L. B., MARCUVITZ, N.: 'Radiation and scattering of waves' (Prentice-Hall, New Jersey, 1973)
2. BACH, H., et al,: 'Modern topics in electromagnetics and antennas' Peter Peregrinus Ltd. Conference publication 13, Stevenage, 1977
3. SILVER, S.: 'Microwave antenna theory and design' (McGraw-Hill, 1949)
4. BOOKER, H. G., and CLEMMOW, P. C.: 'The concept of an angular spectrum of plane waves, and its relationship to that of polar diagram and aperture distribution', *Proc. IEE.*, 1950, **97**, (3), pp. 11–17
5. RUSCH, W. V. T.: 'Physical optics diffraction coefficients for a paraboloid', *Electron. Lett.*, 1974, **10**, pp. 358–360
6. MENTZNER, C. A., PETERS Jun., L., and RUDDUCK, R. C.: 'Slope diffraction and its application to horns', *IEEE Trans*, 1975, **AP-23**, pp. 153–159
7. WOOD, P. J. and BOSWELL, A. G. P.: 'Elliptical beam antenna for satellite applications – Part 1, theory'. IEE International Conference on Antennas for Aircraft and Spacecraft, 3–5 June 1975, pp. 83–88
8. BOWMAN, F.: 'Introduction to Bessel functions' (Dover, New York, 1958)
9. RUSCH, W. V. T.: 'Scattering of a spherical wave by an arbitrary truncated surface of revolution'. JPL Technical Report 32–434, May 1963, California Institute of Technology
10. JAMES, G. L.: 'Geometrical theory of diffraction of electromagnetic waves' (Peter Peregrinus Ltd., 1976)
11. SAFAK,: 'Calculation of radiation patterns of paraboloidal reflectors by high-frequency asymptotic techniques', *Electron. Lett.*, 1976, **12**, pp. 229–231
12. CLAYDON, B., and JAMES, G. L.: 'Asymptotic methods for the prediction of dual-reflector antenna efficiency', *Proc. IEE* 1975, **122**, pp. 1359–1362
13. SOMMERFELD, A. J. W.: 'Optics' (Academic Press, New York, 1954)
14. KELLER, J. B.: 'Geometrical theory of diffraction', *J. Opt. Soc. Am.*, 1962, **52**, pp. 116–130
15. KOUYOUMJIAN, R. G., and PATHAK, P. H.: 'A uniform geometrical theory of diffraction for an edge in a perfectly conducting surface, *Proc. IEEE*, 1974, **62**, pp. 1448–1461
16. RUSCH, W. V. T., and SORENSEN, O.: 'The geometrical theory of diffraction for axially symmetric reflectors', *IEEE Trans.*, 1975, **AP-23**, pp. 414–419
17. JAMES, G. L., and MALIK, D. P. S.: 'Towards the theoretical design of splash-plate feeds', *Electron Lett.*, 1975, **11**, pp. 593–594
18. JENSEN, F., and PORTOPPIDAN, K.: 'V.H.F. radiation patterns of satellite antennas'. IEE International Conference on Antennas for Aircraft and Spacecraft, 3–5 June 1975, pp. 41–46
19. JAMES, G. L., and POULTON, G. T.: 'Modified diffraction coefficients for focusing reflectors', *Electron Lett*, 1973, **9**, pp. 537–538
20. KNOP, C. M., and OSTERTAG, E. L.: 'A note on the asymptotic physical optic solution to the scattered fields from a parabolic reflector', *IEEE Trans.*, 1977, **AP-25**, pp. 531–534

21 BOLOMEY, J. C., and WIRGIN, A.: 'The theoretical determination of the electromagnetic field in the neighbourhood of a parabolic cylinder'. IEE Conf. Publ. 114, URSI Symposium on Electromagnetic Wave Theory, Imperial College London, p. 105

22 ALBERTSEN, N. C., and BALLING, P.: 'Shaped paraboloidal reflectors.: optimisation by minimax and analysis by physical optics and method of moments'. Ticra A/S Report S-46-01, Copenhagen, 1976

23 POGGIO, A.J., and MILLER, E. K.: 'Integral equation solutions for three dimensional scattering problems', *in* 'Mittra, R. (Ed.): 'Computer techniques for electromagnetics' (Pergamon Press, 1973)

24 JONES, D. S.: 'The theory of electromagnetism' (Pergamon Press, New York, 1964)

25 HARRINGTON, R. F.: 'Field computation by moment methods' (Macmillan, New York, 1968)

26 MAUTZ, J. R., and HARRINGTON, R. F.: 'Radiation and scattering from bodies of revolution', *Appl. Sci. Res.*, 1969, **20**, pp. 405–435

27 ALBERTSEN N. C., et al,: 'Computation of spacecraft antenna radiation patterns'. Technical University of Denmark Report R108, Lyngby, 1972

28 KO, W. L., and MITRRA, R.: 'A new approach based on a combination of integral equation and asymptotic techniques for solving electromagnetic scattering problems', *IEEE Trans.*, 1977, **AP-25**, pp. 187–197

Chapter 3

The spherical wave expansion method

3.1 Vector spherical waves

Vector spherical waves are elementary solutions of free-space Maxwell's equations that are separable in spherical polar co-ordinates. At least in the case of the most well known of three possible types of expansion (the Laurent expansion) each elementary wave admits of a simple physical interpretation. Specifically, the wave fields are those of an array of electric or magnetic dipole sources. The array is positioned at the origin of the spherical polar co-ordinate system, and is of limitingly small dimensions. The number of elements relates to the order of the spherical wave. However, virtually all applications of the expansion involve representing a real radiator of finite dimensions by invoking equivalence principles (Section 2.2.4).

The basic formulas for the electromagnetic fields of TE (E suffix) and TM (M suffix) spherical waves are as follows.[1, 2]

$$\boldsymbol{E}_{E,S,m,n} = A \cos m\phi \boldsymbol{a}_\theta - B \sin m\phi \boldsymbol{a}_\phi \tag{3.1}$$

$$\boldsymbol{H}_{E,S,m,n} = j(D \sin m\phi \boldsymbol{a}_\theta + C \cos m\phi \boldsymbol{a}_\phi + E \sin m\phi)\boldsymbol{a}_r \tag{3.2}$$

$$\boldsymbol{E}_{M,S,m,n} = D \cos m\phi \boldsymbol{a}_\theta - C \sin m\phi \boldsymbol{a}_\phi + E \cos m\phi \boldsymbol{a}_r \tag{3.3}$$

$$\boldsymbol{H}_{M,S,m,n} = -j(A \sin m\phi \boldsymbol{a}_\theta + B \cos m\phi \boldsymbol{a}_\phi) \tag{3.4}$$

where

$$A = m z_n(X) P_n^m(Y) \operatorname{cosec} \theta \tag{3.5}$$

$$B = z_n(X) \frac{\partial}{\partial \theta} \{P_n^m(Y)\} \tag{3.6}$$

$$C = \frac{m}{X} \frac{\partial}{\partial X} \{X z_n(X)\} P_n^m(Y) \operatorname{cosec} \theta \tag{3.7}$$

$$D = \frac{1}{X}\frac{\partial}{\partial X}\{Xz_n(X)\}\frac{\partial}{\partial\theta}\{P_n^m(Y)\} \qquad (3.8)$$

$$E = n(n+1)\frac{z_n(X)}{X}P_n^m(Y) \qquad (3.9)$$

and

$$X = kr \qquad (3.10)$$

$$Y = \cos\theta \qquad (3.11)$$

$$z_n(x) = h_n^{(2)}(x) = j_n(x) - jy_n(x) \qquad (3.12)$$

(radially-outward propagating waves)

or

$$z_n(x) = h_n^{(1)}(x) = j_n(x) + jy_n(x) \qquad (3.13)$$

(radially-inward propagating waves)

The dependence on the azimuthal variable ϕ is essentially of Fourier form and either $\cos m\phi/\sin m\phi$ or $\exp(\pm jm\phi)$ formulations are possible. The former option has been selected for this book, and is well adapted to linearly polarised antennas (as opposed to circular polarisation applications).

The subscript 'S' in eqns. 1–4 implies 'symmetric'. Thus an antenna which is linearly polarised with E-vector for $\theta = 0$ parallel to the x-axis and which also has a geometry with mirror image symmetry about this axis can excite only symmetric Fourier components. Moreover, for symmetric spherical waves, the Ludwig definition 3 copolar fields (see Chapter 6) have direct mirror image symmetry about the x-axis. By way of contrast, the crosspolar fields exhibit 'antisymmetry', that is, they are equal in magnitude but opposite in sign.

In general, it is also necessary to include the complementary set of 'asymmetric' spherical waves, identified by an 'A' suffix. This wave set is derived from eqns. 1—4 by rotating the 'ϕ' co-ordinate through $\pi/2m$, that is, by replacing $\cos m\phi$ by $\sin m\phi$ and $\sin m\phi$ by $-\cos m\phi$. For both symmetric and asymmetric waves the expansion indices m, n range as

$$m \leqslant n \leqslant n_L \qquad (3.14)$$

(radial index)

$$0 \leqslant m \leqslant n_L \qquad (3.15)$$

(azimuthal index)

The formulation of the vector spherical wave sets introduces certain special functions via eqns. 5–9, namely associated Legendre functions and spherical Bessel functions. These functions are most conveniently defined by iterative relationships,[1] which also form the basis of an extremely efficient method of numerical evaluation when using a digital computer. Thus for *any* spherical Bessel function ($j_n, h_n^{(1)}$ etc.).

$$z_{n+1} = (2n+1)z_n/x - z_{n-1} \tag{3.16}$$

and the derivative function is obtained via

$$\frac{1}{x}\frac{\partial}{\partial x}(xz_n) = \frac{n+1}{2n+1}z_{n-1} - \frac{n}{2n+1}z_{n+1} \tag{3.17}$$

A forward recursion based on eqn. 16 becomes possible once the $n = 0$ and $n = 1$ functions are specified; for example, for the spherical Hankel function of the second kind,

$$h_0^{(2)}(x) = \frac{\sin x}{x} + j\frac{\cos x}{x} \tag{3.18}$$

$$h_1^{(2)}(x) = \frac{\sin x}{x^2} - \frac{\cos x}{x} + j\left(\frac{\cos x}{x^2} + \frac{\sin x}{x}\right) \tag{3.19}$$

For the associated Legendre functions, the iterative relationships are

$$P_m^m = (2m-1)\sin\theta P_{m-1}^{m-1} \tag{3.20}$$

and

$$(n-m+1)P_{n+1}^m - (2n+1)yP_n^m + (n+m)P_{n-1}^m = 0 \tag{3.21}$$

whereas the derivative functions can be obtained from

$$\sin\theta\frac{\partial}{\partial\theta}(P_n^m) = nyP_n^m - (n+m)P_{n-1}^m \tag{3.22}$$

The starting point for recursion is now specified by

$$P_0^1 = 0 \tag{3.23}$$

and

$$P_1^1 = \sin\theta \tag{3.24}$$

3.2 The Laurent expansion

3.2.1 A source-matched formulation

The Laurent-type expansion* is the most well known of the three fundamental types of spherical wave expansion to be described in this book. It applies to the prototype problem of Fig. 3.1, e.g. prediction

* so called by analogy to the Laurent series in complex variable theory

of radiated field distribution in a region exterior to the 'minimum sphere'. The latter is the sphere of minimum radius $r = R_0$ concentric with the r, θ, ϕ spherical wave co-ordinate system that just encompasses all radiating sources. The radial functions for expansion of the series at an observation point are spherical Hankel functions of the second kind, that is eqn. 12 is applicable.

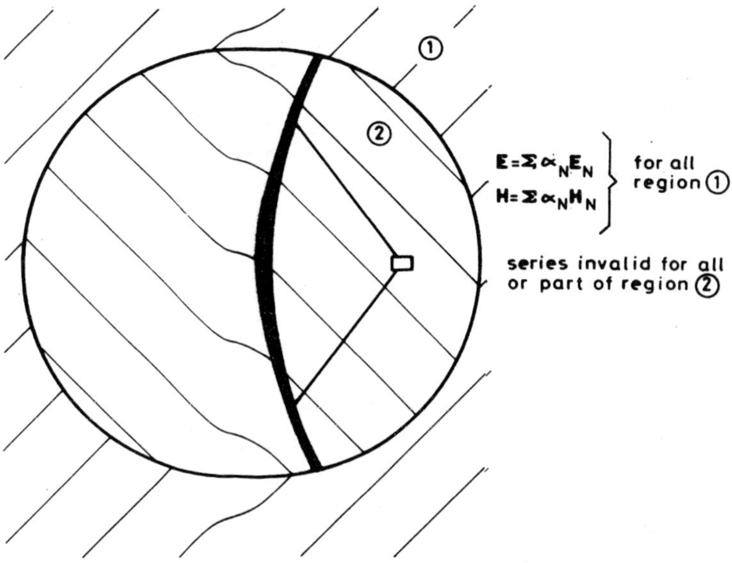

Fig. 3.1. *Laurent expansion*

The first consideration is to determine the spherical wave weighting coefficients $\alpha_{E,S,m,n}$ etc. in terms of the specification of the source. A number of alternative formulations are possible. It is instructive to outline their derivation by a method which invokes the orthogonality properties of a spherical wave set.[2] This if N is a 'bumper' index incorporating

then $\left.\begin{matrix}E\\M\end{matrix}\right\},\left.\begin{matrix}S\\A\end{matrix}\right\}, m, n$

$$\int_S (E_N \times H_{N'} - E_{N'} \times H_N) \cdot dS = 0$$

$$N \neq -N' \tag{3.25}$$

Here N is regarded as a signed quantity, with significance

$N > 0$, Hankel functions of second kind, radially-outward propagation

and

$N < 0$, Hankel functions of first kind, radially-inward propagation

Because of the complex-conjugate relationship between eqn. 12 and eqn. 13

$$E_{-N} = \overset{*}{E}_N \tag{3.26}$$

$$H_{-N} = -\overset{*}{H}_N \tag{3.27}$$

Finally, S in eqn. 25 may be any closed surface encompassing the S.W.E. co-ordinate origin.

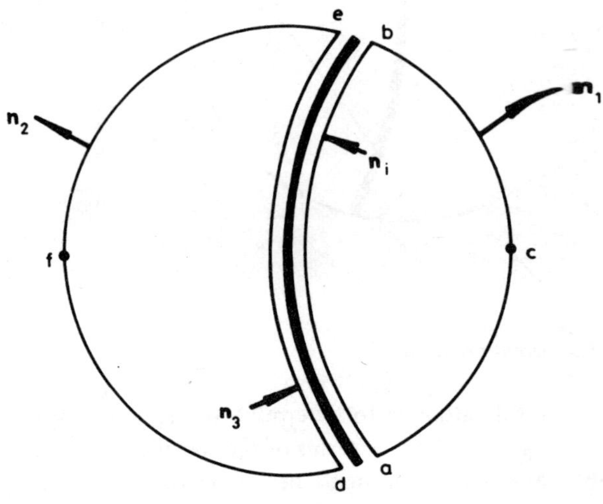

Fig. 3.2. *Integration surfaces, matching SWE to current sheet source*
$S_1 = abc \quad S_i = ab \quad S_2 = dfe \quad S_3 = de$

Now consider the current sheet source of Fig. 3.2. Its radiated fields (in all space) are to be denoted by the distribution $\boldsymbol{E}:\boldsymbol{H}$. The expansion

$$E = \sum_{N=1}^{N=N_{max}} \alpha_N E_N \tag{3.28}$$

$$H = \sum_{N=1}^{N=N_{max}} \alpha_N H_N \tag{3.29}$$

is introduced, valid at least in the region exterior to the minimum sphere $S_1 + S_2$.

The first step to determine α_N is to vector cross-multiply by test fields $\boldsymbol{H}_0$ and $\boldsymbol{E}_0$, respectively, and integrate both sides of eqns. 28 and 29 over $S_1 + S_2$. One choice of test field is

$$\begin{aligned} E_0 &= E_N + E_{-N} \\ H_0 &= H_N + H_{-N} \end{aligned} \tag{3.30}$$

Then, after exploiting the orthogonality relationship of eqn. 25,

$$\alpha_N = \frac{\int_{S_1+S_2} \{E \times (H_N + H_{-N}) + H \times (E_N + E_{-N})\} \cdot dS}{\int_{S_1+S_2} (E_N \times H_{-N} + H_N \times E_{-N}) \cdot dS} \tag{3.31}$$

The denominator may be rewritten using eqns. 26 and 27 as

$$-2 \int_{S_1+S_2} Re[E_N \times \overset{*}{H}_N] \cdot dS = (K_N)^{-1} \tag{3.32}$$

where '*Re*' signifies real part and the integral is therefore recognised as the power flux through $S_1 + S_2$ of the *N*th spherical mode. Thus[1, 2]

$$K_N = -\pi \frac{(2n+1)(n-m)!}{n(n+1)(n+m)!} \tag{3.33}$$

At this point it is expedient to introduce an electromagnetic identity associated with surface integrals of the type which enter in eqn. 31. Thus.

$$I_{AB} = \int_S (E_A \times H_B + H_A \times E_B) \cdot dS \tag{3.34}$$

is identically zero[2] provided that both $\boldsymbol{E}_A : \boldsymbol{H}_A$ and $\boldsymbol{E}_B : \boldsymbol{H}_B$ are analytic solutions of free-space Maxwell's equations within *S*. In fact, these conditions are met for the numerator of eqn. 31 and any surface not intersecting or enclosing the source, in respect of both $\boldsymbol{E}:\boldsymbol{H}$ and $\boldsymbol{E}_N + \boldsymbol{E}_{-N}:\boldsymbol{H}_N + \boldsymbol{H}_{-N}$. In the latter case the particular mode combination selected suppresses the $y_n(kr)$ radial dependence of propagating spherical waves. That is, only the $j_n(kr)$ function is present, and this function is finite and analytic in all space including $r = 0$. Thus referring to Fig. 3.2 we have

$$\int_{S_2+S_3} = \int_{S_i+S_1} = 0 \qquad (3.35)$$

Applying eqn. 35 to eqn. 31 enables the latter to be written as

$$\alpha_N = -K_N \int_{S_i+S_3} (\boldsymbol{E} \times \boldsymbol{H}_{\pm N} + \boldsymbol{H} \times \boldsymbol{E}_{\pm N}) \cdot d\boldsymbol{S} \qquad (3.36)$$

Here an abbreviated '±' notation has been introduced to represent the sum of radially inward and outward waves, e.g. $\boldsymbol{H}_N + \boldsymbol{H}_{-N}$ etc. Because the unit normal vectors implicit in $d\boldsymbol{S}$ are related by

$$\boldsymbol{n}_i = -\boldsymbol{n}_3$$

a contribution to eqn. 36 can only arise in so far as field discontinuities are present. For *an electric current sheet*, the only discontinuity associated with the integrand is that of $\boldsymbol{H}$; specifically from Section 2.2.2.

$$D_{-+}(\boldsymbol{H}) = -\boldsymbol{n}_i \times \boldsymbol{i} \qquad (3.37)$$

Thus eqn. 36 finally reduces to

$$\alpha_N = K_N \int_{S_i} \boldsymbol{i} \cdot \boldsymbol{E}_{\pm N} dA \qquad (3.38)$$

3.2.2 *Alternative formulations*

Alternatively, the derivation may commence by choosing a test field

$$\begin{aligned} \boldsymbol{E}_0 &= \boldsymbol{E}_{-N} \\ \boldsymbol{H}_0 &= \boldsymbol{H}_{-N} \end{aligned} \qquad (3.39)$$

in place of eqn. 30. This results in the formulation

$$\alpha_N = K_N \int_S (\boldsymbol{E} \times \boldsymbol{H}_{-N} + \boldsymbol{H} \times \boldsymbol{E}_{-N}) \cdot d\boldsymbol{S} \qquad (3.40)$$

Although S is initially the minimum sphere $S_1 + S_2$, by invoking eqn. 34 it is possible to expand S to an arbitrary shape. However, it cannot be collapsed to beyond the point where it just encompasses *both* the source and the spherical wave origin. This latter limitation arises because $\boldsymbol{E}_{-N} : \boldsymbol{H}_{-N}$ is not analytic at $r = 0$, which influences the implementation of eqn. 34.

A third choice of test field is

Table 3.1 *Laurent spectrum coefficients for a 16λ paraboloid; m = 1 TM waves for 3 different types of field data (normalised to same power in each case)*

Radial index n	Laurent spectrum derived directly from radiating currents	Laurent spectrum derived by matching to theoretically calculated field distribution on 50λ sphere	Laurent spectrum derived by matching to measured fields on 50λ sphere
1	0·0875	0·0611	0·113
5	0·0214	0·0152	0·0186
10	0·00467	0·00762	0·0129
15	0·00388	0·00489	0·00564
20	0·00486	0·00341	0·00374
25	0·00324	0·00249	0·0256
30	0·00155	0·00191	0·00180
35	0·00102	0·00136	0·00132
40	0·00092	0·00116	0·00096
45	0·00055	0·00088	0·00095
50	0·00028	0·00047	0·00026
55	0·00002	0·00008	0·00004
60	$2{\cdot}6 \times 10^{-6}$	$1{\cdot}1 \times 10^{-5}$	$1{\cdot}1 \times 10^{-5}$
65	$1{\cdot}8 \times 10^{-7}$	$2{\cdot}0 \times 10^{-6}$	$4{\cdot}9 \times 10^{-5}$
70	$2{\cdot}0 \times 10^{-9}$	$3{\cdot}2 \times 10^{-6}$	$4{\cdot}1 \times 10^{-5}$
75	$1{\cdot}5 \times 10^{-10}$	$2{\cdot}9 \times 10^{-6}$	$5{\cdot}2 \times 10^{-5}$
80		$3{\cdot}0 \times 10^{-7}$	$7{\cdot}5 \times 10^{-5}$
85		$7{\cdot}3 \times 10^{-6}$	$1{\cdot}2 \times 10^{-4}$
90	$<10^{-11}$	$3{\cdot}5 \times 10^{-6}$	$1{\cdot}3 \times 10^{-4}$
95		$5{\cdot}2 \times 10^{-6}$	$1{\cdot}1 \times 10^{-4}$
100		$6{\cdot}8 \times 10^{-6}$	$1{\cdot}0 \times 10^{-4}$

(*b*) matching to the secondary wavefront on a 50λ diameter sphere, the fields on this sphere being determined from an analytical model (the model constitutes PO plus calculation of near fields of a 16λ parabolic reflector using Laurent expansion)
(*c*) matching to the secondary wavefront on a 50λ diameter sphere, experimental field data for a real antenna.

For $n > 50$, a rapid convergence sets in for the source-matched case. However, for the secondary wavefront matched spectra, convergence is only realised to the extent that there is a cancellation between 'in-phase' and 'antiphase' contributions to the integrands of eqns. 40 or 43. Thus, the spectrum tends to flatten out to an asymptotic level owing to errors in the data. The corresponding series is only 'partially convergent', and for sufficiently large n will eventually diverge due to the $h_n^{(2)}$ functions of eqns. 28 and 29.

This example will have served to illustrate the advantages of a

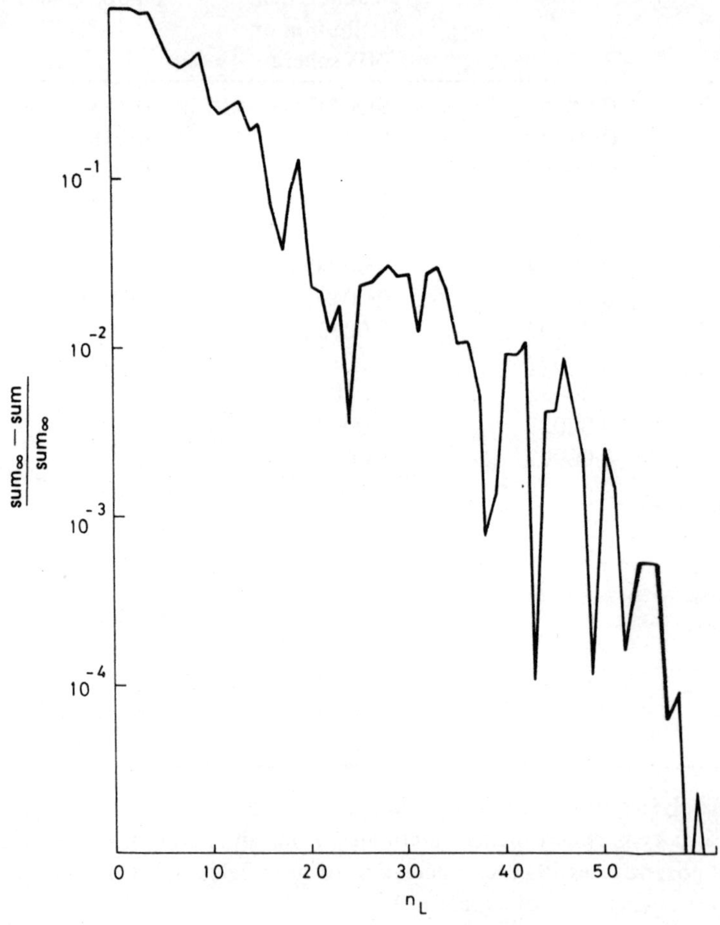

Fig. 3.3. *Convergence of Laurent SWE with radial index*

source-matched representation. As already stated, source-matching always provides a precise, unambiguous solution. It also minimises the numerical working involved in evaluating the weighting coefficients.

Moreover, it lends itself to a simple criterion to select a termination point for the series. Finally, eqn. 38 is 'universal' both in the sense that it transcends all fundamental limitations on the position of the co-ordinate origin for the series, and that it can also accommodate secondary wavefront data if desired. The latter option becomes available when a generalisation of eqn. 38 to include magnetic currents i_M,

$$\alpha_N = K_N \int_{S_{i,\,i_M}} \{i \cdot E_{\pm N} - i_M \cdot H_{\pm N}\} dA \tag{3.46}$$

is adopted and the equivalence formulas of eqn. 2.9 are used to specify i and i_M in terms of the secondary wavefront fields. $S_{i,\,i_M}$ then becomes of necessity a closed surface.

3.2.4 *Criteria for radial index*

An example demonstrating the overall convergence behaviour of a source-matched Laurent series is given in Fig. 3.3. After a certain critical index, corresponding approximately to eqn. 44, the error in the series in relation to its 'sum to infinity' decreases roughly exponentially. The particular case illustrated is the calculation of the scattered pattern of a 13λ hyperboloid subreflector at 30λ range. A maximum radial index n_L of 60 suffices to give an overall accuracy of about one part in 10^5 in this problem.

The origin of the series for Fig. 3.3 is at the hyperboloid vertex. An alternative location at the near focus would introduce a greater degree of coherence between the induced currents and spherical waves, improving somewhat the rate of convergence for small n values.

In general, for a given accuracy n_L is primarily a function of the radius of the outer annulus of current for the source. For large reflector antennas, n_L is essentially proportional to reflector diameter D. However, the relationship is nonlinear for $D \ll 50\lambda$, and also there is some dependence on reflector shape.

Fig. 3.4 gives representative curves for n_L for both a paraboloid and a hyperboloid reflector, corresponding to a one part in 10^5 accuracy criterion which will be adopted as standard in this book. For any given radial index n in eqn. 38, $n < n_L$, only those current annuli having radii greater than r_{min}, where r_{min} is related to n in a similar manner to Fig. 3.4, contribute significantly to the integration. This property can be used to advantage in a numerical evaluation of the spectrum weights. It is illustrated by the 'radial index rings' of Fig. 3.5.

3.2.5 Numerical convergence test

In a numerical implementation of a source-matched S.W.E., it is a simple matter to introduce a test as each term of the series is calculated to see if the convergence point has been reached. A suitable criterion can be stated as follows:[4] terminate the series if the latest two terms are both smaller than 10^{-5} times the latest sum, in respect of all field components. Such a test can be applied during the calculation of the spectrum weights (where it defines an individual truncation point for each annulus of the source, as described in Section 3.2.4). It can also be applied during the re-expansion of the series at an observation point.

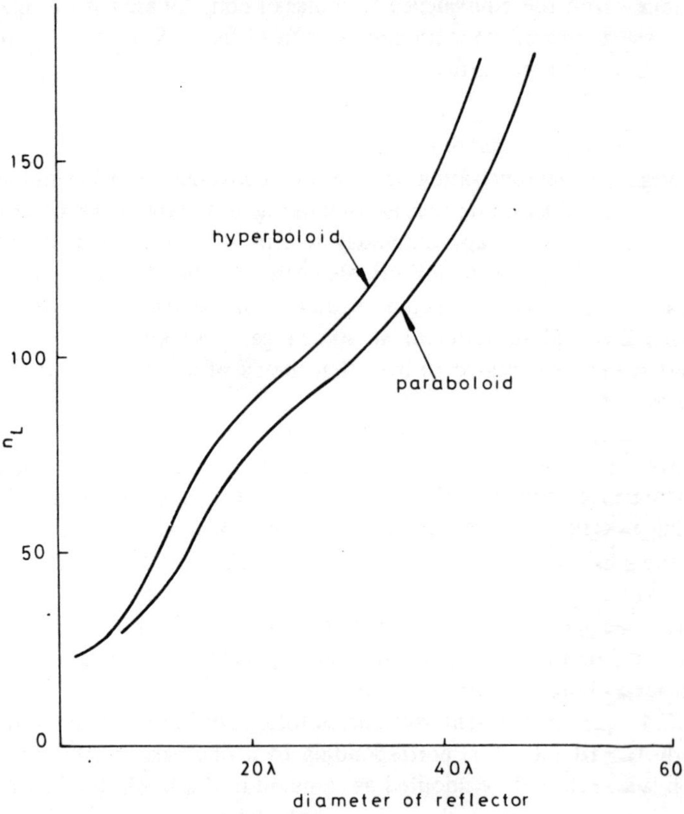

Fig. 3.4. *Maximum radial index n_L for Laurent expansion*

Thus it will be seen that prior knowledge of n_L, e.g. Fig. 3.4, is a requirement only in so far as some estimate of computer time/storage is required in advance of carrying out a computation.

3.2.6 An example: Laurent expansion for offset feed scattering problem

To rephrase the formal relationships of eqns. 28, 29 and 38 as a computing algorithm, it is usually advantageous to reduce them to a more explicit form. The particular case of a surface of revolution Cassegrainian subreflector excited by a non-coaxial feed horn (Fig. 3.6) has been selected as representative of the approach to be adopted.

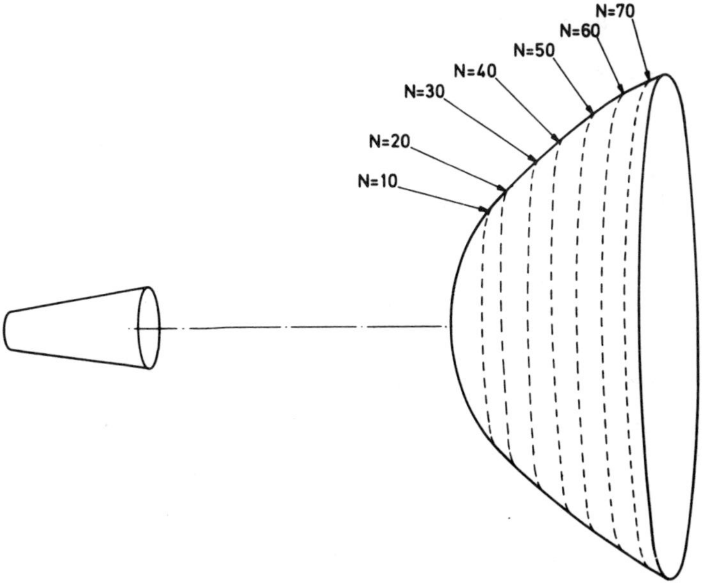

Fig. 3.5. *Illustration of radial index rings for a scatterer*

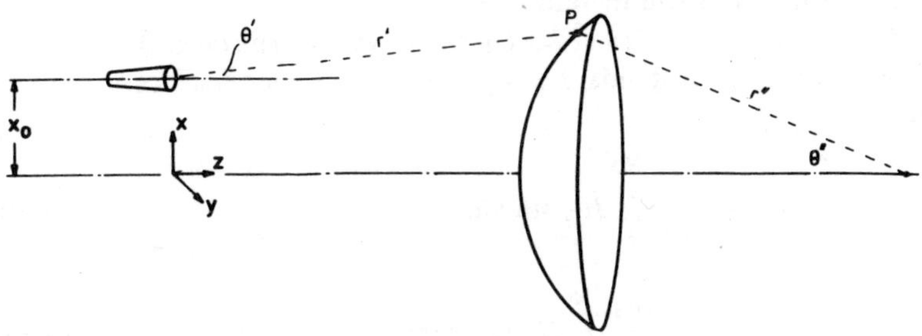

Fig. 3.6. *Problem geometry for an asymmetric scatterer*

The feed horn is assumed to be offset transversely along the x-axis. Its primary radiation in relation to a local co-ordinate system having an origin at the centre of its aperture is assumed to be of the form of eqn. 2.32, with $f_3(\theta)$ zero. Then the unit vectors for the coaxial $r''\theta''\phi''$ system of Fig. 3.6 are related by the geometrical identities

$$\boldsymbol{a}_{\theta'}\cdot\boldsymbol{a}_t = \cos\theta_{norm}\cos\theta'\cos\psi-\sin\theta_{norm}\sin\theta'$$

$$\boldsymbol{a}_{\theta'}\cdot\boldsymbol{a}_{\phi''} = -\cos\theta'\sin\psi$$

$$\boldsymbol{a}_{\phi'}\cdot\boldsymbol{a}_t = -\cos\theta_{norm}\sin\psi$$

$$\boldsymbol{a}_{\phi'}\cdot\boldsymbol{a}_{\phi''} = -\cos\psi$$

where

$$\psi = \phi'+\phi''$$

Here $\boldsymbol{a}_t$ is a unit vector orthogonal to both the reflector normal $\boldsymbol{n}$ and $\boldsymbol{a}_{\phi}''$. Eqn. 2.32 then transforms to

$$H_{ft} = (\cos\theta_{norm}\cos\theta'\cos\psi\sin\phi'-\sin\theta_{norm}\sin\theta'\sin\phi')f_1(\theta')$$
$$-\cos\theta_{norm}\sin\psi\cos\phi' f_2(\theta') \tag{3.47}$$

and

$$H_{f\phi''} = -\cos\theta'\sin\psi\sin\phi' f_1(\theta')-\cos\psi\cos\phi' f_2(\theta') \tag{3.48}$$

We can now expand H_f in a Fourier series according to

$$\boldsymbol{H}_f = \sum_m ({}_{1S}f_m \sin m\phi'' + {}_{1A}f_m \cos m\phi'')\boldsymbol{a}_t$$
$$+({}_{2S}f_m\cos m\phi''-{}_{2A}f_m\sin m\phi'')\boldsymbol{a}_{\phi''} \tag{3.49}$$

The suffices S, A above correspond to the convention adopted in Section 3.1. In the present case, the asymmetric components are identically zero because of the mirror image symmetry about the x-axis. The A suffix terms will therefore be omitted forthwith.

Suppose $\boldsymbol{H}_f$ is specified at N_ϕ equal intervals in the range $0 \leqslant \phi'' < 2\pi$. Then, according to standard sampled-data Fourier transform relationships,

$${}_{1S}f_m = \frac{2}{N_\phi}\sum_1^{N_\phi} H_{ft}\sin m\phi'' \tag{3.50}$$

$${}_{2S}f_m = \frac{2\delta(m)}{N_\phi}\sum_1^{N_\phi} H_{f\phi''}\cos m\phi'' \tag{3.51}$$

Table 3.2 *SWE versus KHS computational requirements: a numerical example*

Example: 25λ circularly symmetric subreflector, cos ϕ/sin ϕ excitation
Tabulation of scattered patterns: 0°–180°, 1° ints. in E- and H-planes
Accuracy requirement: 1 pt. in 10^{-4}
No of integration points: 250

KHS method (far field)
8 'trig' functions per integration point × 250 integration points × 180 pattern points (360 000 'trig' functions)

SWE method
Coefficients: 60 arithmetic operations × 105 radial indices × 250 integration points (≃ 787 000 arithmetic functions)
re-expansion: 60 arithmetic operations × 105 radial indices × 180 pattern points (11 34 000 arithmetic functions)

Table 3.3 *TM Laurent spectral coefficients for a focused reflector*

Radial index n	151 integration points	15 integration points
1	0·233 ∠ 93·3°	0·157 ∠ 350·6°
5	0·0498 ∠ 70·4°	0·0463 ∠ 101·4°
10	0·0233 ∠ 204·7°	0·0228 ∠ 229·8°
15	0·0140 ∠ 302·2°	0·0188 ∠ 286·0°
20	0·0119 ∠ 345·0°	0·0146 ∠ 338°
25	0·00125 ∠ 113·6°	0·00375 ∠ 136·9°
30	0·00252 ∠ 171·7°	0·00259 ∠ 163·9°
35	0·00148 ∠ 345·5°	0·00244 ∠ 190·8°
40	0·000943 ∠ 345·5°	0·000418 ∠ 303·4°
45	0·000338 ∠ 93·9°	0·000915 ∠ 15·0°
50	0·000306 ∠ 248·0°	0·000407 ∠ 200·2°
55	0·0000546 ∠ 66·8°	0·0000812 ∠ 10·5°
60	0·00000186 ∠ 266·0°	0·0000102 ∠ 182·4°
65	0·00000029 ∠ 40·8°	0·00000044 ∠ 0·6°

3.2.9 Focused apertures

Table 3.3 illustrates the Laurent spectrum for a 15λ diameter circularly symmetric paraboloid. Only every fifth coefficient has been included for brevity. For the first half of the table, the spectrum was calculated by a numerical integration using 151 points between middle and edge of the reflector. This number is just sufficient to meet criteria for convergence of Simpson's numerical quadrature rule for an accuracy of one part in 10^5 (Section 3.5.4). However, it represents far in excess of that number which would ordinarily be needed when implementing the KHS method to calculate the 'near-in' (main lobe and first few sidelobes) radiation pattern for the same problem.

For the second part of the table, the number of integration points has been reduced to 15. This change is seen to result in a completely different set of spectrum coefficients.The power content of the latter has increased in relation to that of the original spectrum by a 1·38 factor. However, when the series is re-expanded to calculate the far field, either spherical wave spectrum yields a virtually identical near-in pattern (see Table 3.4). The 15-point integration grid spectrum does, however, yield a different (incorrect) wide angle pattern. The far out sidelobes are at an unrealistically high level.

Table 3.4 *Comparison of patterns using Laurent series for a focused reflector (H-plane)*

Angle off-axis	151 integration point spectral coefficients		15 integration point spectral coefficients	
	Magnetic field amplitude (voltage scale)	phase (degrees)	Magnetic field amplitude (voltage scale)	phase (degrees)
0°	10·60869	233·1	10·61898	233·1
2°	8·35611	238·9	8·35621	238·9
4°	4·20456	263·6	4·20453	263·6
6°	1·97082	−46·4	1·97142	−46·4
8°	0·72911	3·9	0·72880	3·8

This example serves as a simple illustration of the 'redundancy' properties of the Laurent spectrum for an antenna of the focused aperture type. There are many linearly independent spectrum sets, all representing the same 'near-in' pattern data. A degree of correlation may be discerned between neighbouring spectral components. Specifically, 'groups of 3' may be recognised for which two approximately equal and antiphase components bracket a 'null' component of smaller amplitude. Also the amplitude of the spectral lines tends to define two or more approximately smooth curves. The latter phenomenon is illustrated by Fig. 3.8 for the case of a 16λ paraboloid.

The extent of the redundancy increases with the size of the focused aperture. At the same time, the S.W.E. tends to be more at a disadvantage in numerical terms. Typically, the 'break-even' point in comparison to the KHS method is a diameter ≈ 100λ. To overcome this limitation of the S.W.E., it would be necessary to consider 'thinning' the spectrum in some manner.

3.3 The Taylor expansion

3.3.1 Prototype problem and formulation

Some antenna problems are of a type involving the radiation of an exterior source into an interior region (Fig. 3.9.). The Taylor-type

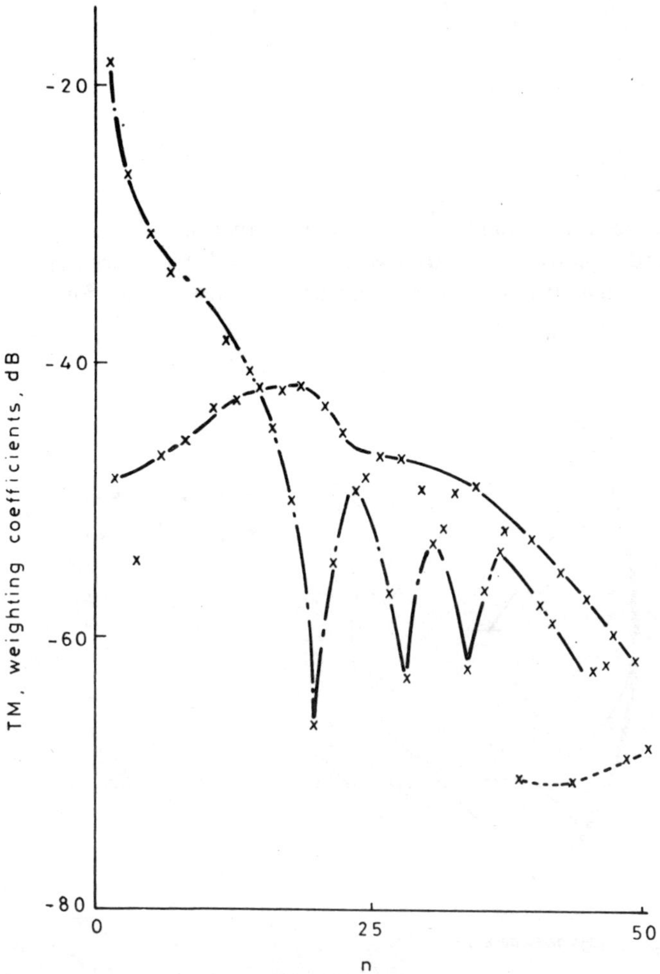

Fig. 3.8. *SWE spectrum for a focused reflector*

expansion* applicable to this case may be specified directly in terms of the radiating current distribution $\boldsymbol{i}$ present on an open or closed surface S_i. In this respect the expansion is similar to the source-matched Laurent series already described. However, the appropriate formulation is now.[2]

$$E = \sum_{N=1}^{N=N_{max}} E_{\pm N} \tag{3.66}$$

$$H = \sum_{N=1}^{N=N_{max}} H_{\pm N} \tag{3.67}$$

with spectrum coefficients specified by

$$\alpha_N = K_N \int_{S_i} i \cdot E_N dA \tag{3.68}$$

The region for validity of the Taylor expansion is delimited by the 'maximum sphere'. The latter is the *largest* sphere concentric with the S.W.E. origin that does not intercept any part of the source current distribution (Fig. 3.9).

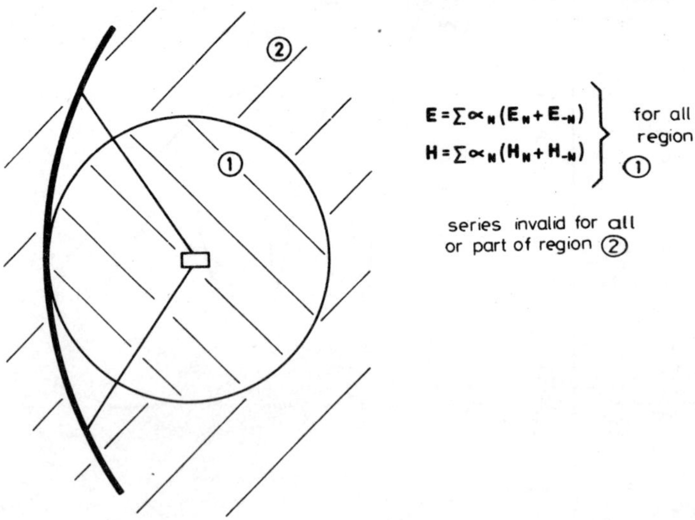

Fig. 3.9. *Taylor expansion*

* So-called by analogy to the Taylor series in complex variable theory.

3.3.2 Radial functions and convergence properties

The radial functions are now $j_n(x')$ for expanding the terms of the series (eqns. 66 and 67), and $h_n^{(2)}(x)$ for the evaluation of the spectrum weights (eqn. 68). Thus the latter now diverge without limit when n is increased through values larger than x. Table 3.5 illustrates this behaviour, for a Taylor series appropriate to focal plane analysis for a 25λ diameter paraboloid (a detailed discussion of the formulation of this problem is given in Reference 2).

Table 3.5 *Typical spectrum coefficients for Taylor series*

n	$TE_{S,1,n}$ and $TE_{S,1,-n}$ coefficient		$TM_{S,1,n}$ and $TM_{S,1,-n}$ coefficient	
	Real	Imaginary	Real	Imaginary
1	1·057	0·772	0·781	−1·050
5	0·0211	0·0113	0·0098	−0·0217
10	0·00027	0·00781	0·00769	−0·00040
15	−0·00172	0·00219	0·0196	0·00184
20	−0·000524	0·000413	0·000111	0·000621
25	0·000263	0·000038	−0·000182	−0·000168
30	0·000445	−0·000206	−0·000266	−0·000335
35	0·000178	−0·000376	−0·000302	−0·000161
40	−0·000141	−0·000175	−0·000172	0·000027
45	−0·000011	0·000066	−0·000060	0·000023
50	0·000020	−0·000120	−0·0000561	0·000064
55	−0·000128	0·000096	0·000067	0·000052
60	0·000129	−0·000035	0·000017	−0·000055
65	−0·000067	0·000064	−0·000047	−0·000060
70	−0·00057	0·00155	0·00065	−0·00182
75	−0·083	0·257	0·107	−0·327
80	−20·7	63·4	29·2	−89·0
85	$-7{\cdot}5 \times 10^3$	$2{\cdot}29 \times 10^4$	$1{\cdot}16 \times 10^4$	$-3{\cdot}52 \times 10^4$
90	$-3{\cdot}9 \times 10^6$	$1{\cdot}18 \times 10^7$	$6{\cdot}4 \times 10^6$	$-1{\cdot}95 \times 10^7$
95	$-2{\cdot}8 \times 10^9$	$8{\cdot}4 \times 10^9$	$4{\cdot}9 \times 10^9$	$-1{\cdot}49 \times 10^{10}$
100	$-2{\cdot}6 \times 10^{12}$	$8{\cdot}0 \times 10^{12}$	$5{\cdot}0 \times 10^{12}$	$-0{\cdot}53 \times 10^{13}$

Since the observation point is now 'internal', $x' < x$. Thus, because of eqn. 45, the $j_n(x')$ functions of the *terms* force rapid overall convergence of the series. In fact, the convergence properties are in many respects analogous to those of the source-matched Laurent series, being unconditional for any current distribution, and maintained in practice for observation points at a radius only a few per cent smaller than that of the maximum sphere.

On re-expansion, a suitable termination point for the Taylor series may be selected by applying the numerical test of Section 3.2.5. However (and in contrast to the Laurent series), at the initial stage of calculating spectrum weights it is not possible to apply such a test. Therefore it is necessary to make some prior, conservative estimate of the largest radial index spectrum coefficient that might be called up on re-expansion. The empirical estimate,

$$n_L = 1{\cdot}5\,kr_{max} \tag{3.69}$$

may be used for this purpose. Here r_{max} is the displacement of the most distant observation point from the S.W.E. co-ordinate origin.

3.3.3 Coherent excitation

The Taylor S.W.E. is frequently used to calculate fields in the vicinity of a GO focus point. This point then constitutes the natural choice of location for the origin of the expansion. The phase variation implicit in the $h_n^{(2)}$ function of the integrand of eqn. 68 is related to that of the complex exponential kernel of eqns. 18 and 19. The latter in turn approximately corresponds to the conjugate of the phase term for the source currents $\boldsymbol{i}$.

As a result, low-radial index waves are of greater amplitude and make a more significant contribution towards the total vector at an observation point than for the usual Laurent expansion. Typically, it is possible to terminate the series at (say) $n = 1/3n_L$ and still obtain a rough approximation to the 'sum to infinity'.

An example of the calculation of fields near a GO focus will now be given.

3.3.4 Example: Taylor expansion for offset reflector focal plane

Fig. 3.10 depicts an offset paraboloid receiving antenna. The reflector surface is that defined by the intersection of a parabola of revolution of focal length F with a circular cylinder of diameter D. The cylinder axis is separated by a displacement d_0 from the axis of revolution. The plane wave excitation is supposed to be incident axially, although a relatively minor generalisation would treat the off-axis case.

Much of the analysis of Section 3.2.6 for a rotationally symmetric reflector is relevant, despite the entirely different problem geometry in the present case. However, the co-ordinate system r, θ, ϕ for the S.W.E. is now advantageously centred on the geometrical focus point (0 of

Fig. 3.10), with the $+z$-axis drawn towards the reflector, and the x-axis parallel to the plane of offset. Moreover, in relation to Section 3.2.6:

(*a*) The auxiliary $r''\,\theta''\,\phi''$ system is now dispensed with: instead, the incident excitation is represented in the r, θ, ϕ system, so that eqn. 49 is replaced by

$$H_{pw} = -\sum_m \left\{ {}_1sf_m \sin m\phi \mathbf{a}_t + {}_2sf_m \cos m\phi \mathbf{a}_\phi \right\} \tag{3.70}$$

(*b*) Eqns. 53–59 are still pertinent, but the normal angle θ_{norm} is now $-\theta/2$. The minus sign arises because fields are incident on a concave (as opposed to convex) surface. The integrand radial functions are now spherical Hankel functions so that the '±' suffix of eqns. 58–59 is no longer applicable.

(*c*) Lower case symbols a, b, c, d, e denote that the exp $(-jx)$ factor for the initialisation of the recursion for $h_n^{(2)}$ is now dropped from eqns. 18 and 19, in anticipation of a cancellation of this phase term with that of the incident plane wave.

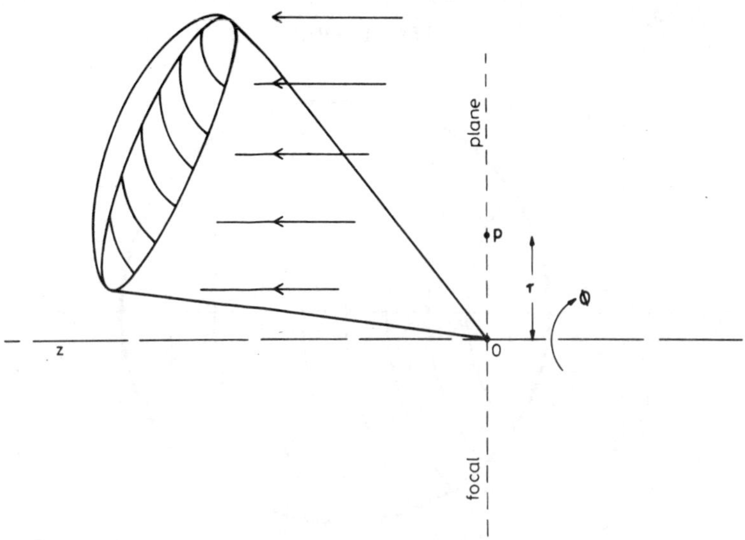

Fig. 3.10. *Formation of focal plane pattern with offset reflector*

The spectrum weights for the offset reflector are then given by

$$\alpha_N = \frac{2\pi K_N}{\delta(m)} \int_0^{\theta_{max}} I_{\alpha_N} r^2 \sin\theta \sec\theta/2 \, d\theta \tag{3.71}$$

where

$$(I_{\alpha_N})_{E,S} = -a\cos\theta/2\ {}_{2S}f_m - b_{1S}f_m \tag{3.72}$$

and

$$(I_{\alpha_N})_{M,S} = -(d\cos\theta/2 + e\sin\theta/2)_{2S}f_m - c\ {}_{1S}f_m \tag{3.73}$$

For an incident plane wave polarised parallel to the x-axis and having unit power flux over the area intercepted by the reflector

$$H_{pw} = \frac{2}{\sqrt{\pi}D}(\cos\theta/2\sin\phi \boldsymbol{a}_t + \cos\phi \boldsymbol{a}_\phi)$$

and its Fourier coefficients are

$$\begin{aligned} {}_{1S}f_m &= -\frac{2\cos\theta/2}{\pi^{3/2}D}\int_{-\phi_L}^{\phi_L}\sin\phi\sin m\phi d\phi \\ \text{and} \\ {}_{2S}f_m &= -\frac{2\delta(m)}{\pi^{3/2}D}\int_{-\phi_L}^{\phi_L}\cos\phi\cos m\phi d\phi \end{aligned} \tag{3.74}$$

Here ϕ_L is the value of ϕ defining the rim of the reflector in the r, θ, ϕ system (see Fig. 3.11),

$$\phi_L = \cos^{-1}\left(\frac{d_0^2 + 4F^2\tan^2\theta/2 - D^2/4}{4Fd_0\tan\theta/2}\right) \tag{3.75}$$

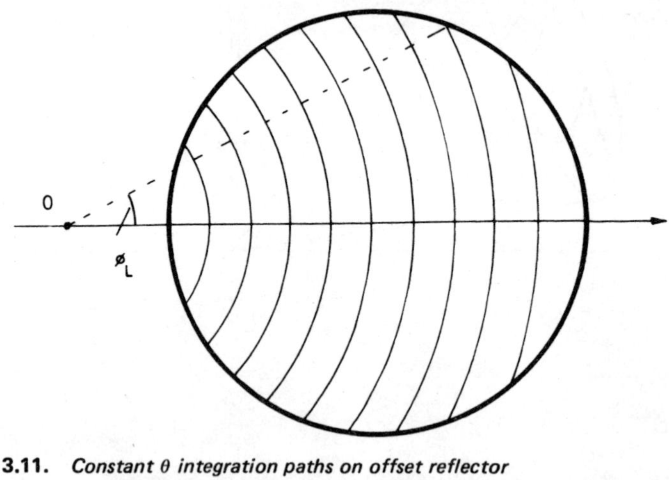

Fig. 3.11. *Constant θ integration paths on offset reflector* view projected onto a plane ⊥ to axis of revolution through O

The integrals of eqn. 74 can readily be evaluated analytically: eqn. 71 can then be evaluated by a numerical quadrature integration rule. The sharp truncation of 'constant θ' annuli at $\phi = \pm\phi_L$ results in

Fourier spectra ${}_{1S}f_m$ and ${}_{2S}f_m$ which decay only gradually with m. Nevertheless, the radial transformation (specifically eqn. 71) ultimately enforces a rapid overall convergence.

The series may be re-expanded according to eqns. 66 and 67 to yield the field anywhere within the maximum sphere of radius F. For evaluation of fields close to the geometrical focus, very few terms are necessary. For example, Figs. 3.12 and 13 appertain to a 20λ diameter reflector of total opening angle 90°. The axis of revolution is offset just to the reflector rim. The contour plots represent fields in a plane through the focus that is perpendicular to the axis of revolution. They were generated with $n_L = 15$, $m_L = 6$ for an accuracy of about one part in 10.[4]

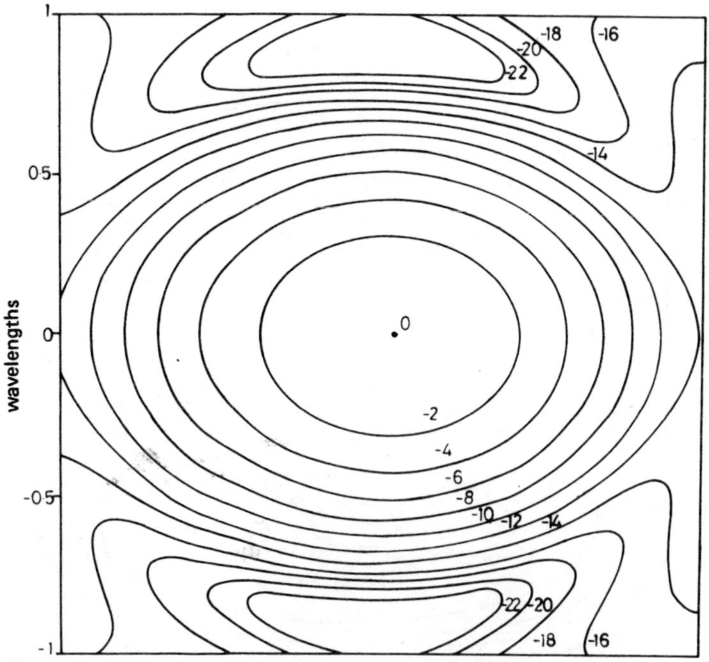

Fig. 3.12. *Copolar field contours in focal plane of offset reflector*

In fact the Taylor solution is extremely efficient in numerical terms in comparison to a KHS vector diffraction treatment. This is in contrast to the behaviour of the Laurent series for a (far-field) focused aperture problem (Section 3.2.9).

3.4 The Uniform Expansion

3.4.1 Derivation

The Taylor and Laurent expansions of the previous sections are suited to applications involving calculation of fields radiated by specified sources. However, in a problem of the 'inverse' type, the specified quantity is the field distribution, and it is required to determine the source currents. To accomplish this type of transformation via an expansion, it is a necessary feature that the latter is valid in all regions of space. In particular, it must remain valid and convergent at the surface of the source. An expansion which meets these requirements will now be described.

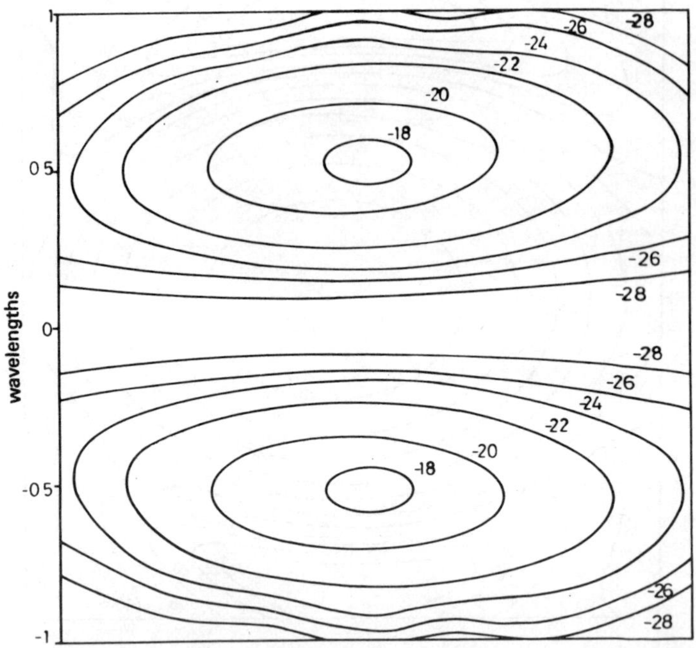

Fig. 3.13. *Crosspolar field contours in focal plane of offset reflector*

The starting point for the derivation of the Uniform expansion is the Laurent S.W.E. of Section 3.2. The special significance of a cophasal current distribution appears first to have been appreciated (in a different context to that of the present application) by Harrington.[5] Let the radiated fields of the source at an arbitrary point in space be denoted by

$$E = E_{ip} + jE_{pq}$$

$$H = H_{ip} + jH_{pq}$$

In addition to the cophasal conditions, suppose that the time reference is selected so that the source current distribution is a real quantity, as denoted by the superscript 'R'. Then from eqn. 38

$$E_{ip} = \frac{1}{2}(E + \overset{*}{E}) = \frac{1}{2} \sum_{N=1}^{N=N_{max}} \left\{ K_N \int_{S_i} i^R \cdot E_{\pm N} dA \right\} E_{\pm N} \qquad (3.76)$$

and

$$jH_{pq} = \frac{1}{2}(H - \overset{*}{H}) = \frac{1}{2} \sum_{N=1}^{N=N_{max}} \left\{ K_N \int_{S_i} i^R \cdot E_{\pm N} dA \right\} H_{\pm N} \qquad (3.77)$$

It may readily be demonstrated that if $E:H$ is an analytic solution of free-space Maxwell's equations, so also is $E_{ip}:jH_{pq}$. The validity of the expansion for the latter given by eqns. 76 and 77 appears at first sight to be that of the parent Laurent expansion, e.g. a minimum sphere restriction. However,

(*a*) The physical field $E:H$ suffers discontinuities across a current sheet source as specified by eqns. 2.6–2.8. However, for cophasal currents the latter equations *specifically exclude* electric field in phase with i, and magnetic field in phase quadrature with i. That is, the physical field $E_{ip}:jH_{pq}$ is analytic at all points on and within the minimum sphere.

(*b*) The expansion on the RHS of eqns. 76 and 77 constitutes a finite sum in which the only radial functions involved are 'j_n' functions. Since the latter are analytic in all space, so is the expansion.

The two statements above taken together are sufficient to prove that eqns. 76 and 77 represent the physical fields *in all space*, by application of the principle of analytic continuation.[6]

A parallel derivation for a magnetic current sheet yields the equations

$$jE_{pq} = \frac{1}{2}(E - \overset{*}{E}) = -\frac{1}{2} \sum_{N=1}^{N=N_{max}} \left\{ K_N \int_{S_{i_M}} i_M \cdot H_{\pm N} dA \right\} E_{\pm N}$$

(3.78)

$$H_{ip} = \frac{1}{2}(H + \overset{*}{H}) = -\frac{1}{2}\sum_{N=1}^{N=N_{max}} \left\{ K_N \int_{S_{i_M}} i_M \cdot H_{\pm N}\, dA \right\} H_{\pm N} \tag{3.79}$$

As for eqns. 76 and 77, the magnetic Uniform expansion represents a cophasal field distribution of essentially 'standing wave' character.

3.4.2 *Physical optics of cophasal currents*

The quantity $E_{ip} : jH_{pq}$ represents a 'partial field', or specifically a component of $E : H$ which is continuous across a current sheet source. It may be evaluated at the source surface S_i, but does not contain sufficient information to specify i in the most general case. Indeed, it would be very surprising if it did, since this would imply that i could be determined uniquely from knowledge of a far field pattern alone. However, i can be specified uniquely if the source surface is specified, *and* an appropriate constraint is put to suppress distributions associated with large reactive energy, e.g. 'supergain' solutions. In the present case, the constraint will be provided by introducing the physical optics principle.

To the order of PO, for electric current sources the tangential component of E_{ip} and the normal component of jH_{pq} *at the surface of the source* do not constitute a partial field, but rather represent the total field. The linearity properties of eqns. 76 and 77 then extend the solution to the case of arbitrarily phased currents. Thus

$$(E_t)_{S_i} \approx \frac{1}{2}\sum_{N=1}^{N=N_{max}} \alpha_N (E_{\pm N} \cdot a_t) \tag{3.80}$$

and

$$(H_{norm})_{S_i} \approx \frac{1}{2}\sum_{N=1}^{N=N_{max}} \alpha_N (H_{\pm N} \cdot a_{norm}) \tag{3.81}$$

Here the spectrum weights α_N are those of the parent Laurent series. Thus they can be determined by matching to exterior or far field data, using eqns. 40, 43 and 46. Stated simply, eqns. 80 and 81 imply that the regular Laurent series may be used to evaluate certain specific field components at the surface of the source merely by dropping the Neumann function that would ordinarily be incorporated on re-expansion of the series.

For the dual case of a magnetic current radiator,

$$(E_{norm})_{S_{i_M}} = \frac{1}{2}\sum_{N=1}^{N=N_{max}} \alpha_N (E_{\pm N} \cdot a_{norm}) \tag{3.82}$$

and

$$(H_t)_{S_{iM}} = \frac{1}{2} \sum_{N=1}^{N=N_{max}} \alpha_N(\boldsymbol{H}_{\pm N} \cdot \boldsymbol{a}_t) \tag{3.83}$$

In principle, a knowledge of E_t or H_t over the (open) surface S_i or S_{iM} permits a rigorous solution for $\boldsymbol{i}$ or i_M via matrix methods (Section 2.4.1.). A more explicit formulation can be obtained at the cost of some further approximation by assuming that the source is a scatterer for which the incident excitation electric and magnetic fields are related at the surface as a plane wave. Thus the PO principle of Section 2.2.6 becomes

$$\boldsymbol{i} \cdot \boldsymbol{a}_u = 2(\boldsymbol{E}_{inc} \cdot \boldsymbol{a}_u) \sec \psi_u$$

$$\boldsymbol{i} \cdot \boldsymbol{a}_v = 2(\boldsymbol{E}_{inc} \cdot \boldsymbol{a}_v) \cos \psi_u$$

Here the unit vector $\boldsymbol{a}_u$ is defined by the intersection of the surface and the plane of incidence, $\boldsymbol{a}_v$ is the orthogonal in-surface vector, and ψ_u is the angle of incidence measured from the surface normal. For an electric current sheet scatterer, the radiated electric field is just the negative of the incident electric field to ensure boundary conditions are satisfied. Thus in terms of the fields of eqns. 80 and 83,

$$\boldsymbol{i} \cdot \boldsymbol{a}_u = -2(\boldsymbol{E}_t \cdot \boldsymbol{a}_u) \sec \psi_u \tag{3.84}$$

$$\boldsymbol{i} \cdot \boldsymbol{a}_v = -2(\boldsymbol{E}_t \cdot \boldsymbol{a}_v) \cos \psi_u \tag{3.85}$$

whereas

$$\boldsymbol{i}_M \cdot \boldsymbol{a}_u = -2(\boldsymbol{H}_t \cdot \boldsymbol{a}_u) \sec \psi_u \tag{3.86}$$

$$\boldsymbol{i}_M \cdot \boldsymbol{a}_v = -2(\boldsymbol{H}_t \cdot \boldsymbol{a}_v) \cos \psi_u \tag{3.87}$$

3.4.3 Application to inverse problems

3.4.3.1 Real and 'equivalent' sources: The prototype inverse problem demands that the antenna radiating currents be derived from analytical or measured data pertaining to its exterior fields. *If*

(*a*) the latter data is specified at and accurate for all θ, ϕ and
(*b*) eqn. 80 can be evaluated at the precise surface S_i at which the physical currents actually reside

then the Uniform expansion echoes physical reality in that eqn. 83 yields a very small value for $\boldsymbol{H}_t$, and hence for the magnetic current distribution. In fact, the calculated value of i_M will then be non-zero only in consequence of the use of the PO approximation. Moreover, eqn. 82 will also yield a very small value for E_{norm}; that is, the Uniform expansion E-vector lies at all points within a single surface, defining the location of the source.

In practice conditions (*a*) and (*b*) are frequently not satisfied. For example, measured far field data is often available only for the 'forward hemisphere' (or a yet more restricted region of space). It is a fundamental property of electromagnetic fields that the distinction between electric and magnetic current sources only becomes evident when rear hemisphere fields are also specified (Section 2.2.5).

Condition (*b*) will ordinarily be contravened in the case of a reflector antenna with feed support structure. The latter represents a scatterer to which the PO principle cannot be applied. This is the case both because of self-shadowing/multiple reflection effects arising in conjunction with the reflector, and also often because the cross-section of a support strut is not sufficiently large in terms of the wavelength. Thus, it is not meaningful to evaluate the Uniform expansion on the surface of the feed support structure. Rather, the expansion can be evaluated for a surface corresponding to that of the reflector only. The source currents calculated for this surface then include 'equivalent strut' contributions.

3.4.3.2 Example: These considerations may be further amplified by introducing a concrete example. The current distribution i_0 represented by the broken curve of Fig. 3.14. is assumed to reside on a 16·6λ parabolic reflector. The Laurent expansion of eqns. 38, 28 and 29 is first used to calculate the fields radiated by i_0 on a sphere of 50λ radius surrounding the antenna. A new Laurent series is then matched to the spherical surface field data, using eqn. 46. The integration limits in the latter equation are chosen to correspond to a cone semiangle θ_{max} of 44°; this is equivalent to assuming the radiated fields are zero for $\theta > 44°$. Finally, eqns. 80–87 are used to compute the source distributions i and i_M on a parabolic surface.

Numerically, i and i_M are found to be nearly equal. *Twice* i is plotted for comparison with i_0 in Fig. 3.14. Some minor discrepancies are evident, including a small but finite predicted current on the parabolic surface beyond the rim of the reflector. However, this is compatible with known properties of the PO approximation. If the calculation is repeated for $\theta_{max} = 180°$ (full sphere data), the computed electric currents approximately double in magnitude, and are now directly comparable with i_0. The computed magnetic currents however are of much smaller magnitude, a suppression of 30 dB being typical.

3.4.3.3 Comparison with plane wave spectrum inverse transform: The plane wave spectrum of Section 2.2.5 also possesses an inverse form. Moreover, the PWS has an in-built capability for suppressing 'supergain' solutions. This is accomplished simply by omitting the evanescent mode (Section 2.2.5.2) part of the spectrum. The effect is to limit the reactive energy associated with the near fields for the Uniform expansion, a corresponding constraint is associated with the calculation of a tangential electric field *in-phase* with the source current.

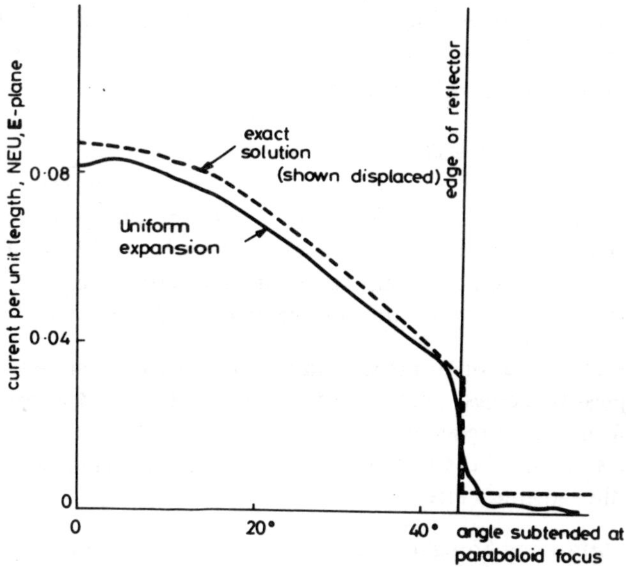

Fig. 3.14. *Prediction of current distribution for a parabolic reflector*

Here, however, the analogy ends. The Uniform expansion removes explicitly the 'minimum sphere' restriction of the basic Laurent S.W.E. However, the propagating wave PWS is still subject to the same 'minimum plane' restriction as the basic PWS. *Thus, in the PWS case, it is not meaningful to evaluate the series at the surface of the source* unless the latter lies entirely in the $z = 0$ plane.

Thus, the salient advantage of the Uniform expansion lies in its ability to deal with the nonplanar source distributions encountered in real reflector antenna applications. Specifically,

(*a*) the source surface may have any arbitrary shape
(*b*) the 'observation surface' on which the 'exterior field data' is specified may have any arbitrary shape
(*c*) the 'exterior field data' itself may have arbitrary properties.

A solution to the inverse problem using the Uniform expansion is most readily achieved when rotational symmetry is present for surfaces (*a*) and (*b*), together with pseudo-circular symmetry for the field data. An S.W.E. with $m = 1$ is naturally adapted to this type of application (Section 3.4.4.3). However, in numerical terms, the technique remains quite feasible and efficient when these symmetries cannot be invoked. Because only the j_n radial functions are involved in the Uniform expansion, the numerical working is largely concerned with arithmetic operations with real variables.

3.4.4 *The restricted synthesis problem*

3.4.4.1 Ideal and realisable patterns: The synthesis problem is set apart from the inverse problem of 3.4.3 in as much as its starting point is an 'ideal' radiation pattern. The latter is usually theoretically unrealisable in the sense that to achieve it precisely would require either an infinite aperture *D* or a supergain antenna. Thus in general pattern synthesis for a reflector antenna involves two fundamental processes:

(*a*) The determination of the spectrum weights of the realisable pattern that represent the best possible approximation to the ideal pattern for the available size of reflector.
(*b*) The solution of the inverse problem for the realisable pattern; this defines the radiating currents.

Step (*a*) can be brought about by truncating a Laurent spectrum for the ideal pattern. Specifically,

$$n_L = 3{\cdot}4D/\lambda \tag{3.88}$$

is found to give an appropriate truncation point for the radial index of the expansion. This truncation process may be thought of as partially suppressing the 'supergain' properties of the ideal pattern with respect to an available aperture *D*. Strictly, however, a truncated S.W.E. is always inherently supergain to a limited extent.*

A penalty associated with this simple technique for calculating a realisable pattern is that the approximation involved causes a 'spread' of the calculated current distribution beyond the intended aperture. This occurs to a greater extent than that due to the PO approximation in the inverse problem.

* For such a series the usual minimum sphere restriction is not applicable, and re-expansion within the sphere demonstrates the presence of large reactive fields.

where

$$(I_{\alpha_N})_{E,S} = -(B_\pm - jD_\pm \cos\psi - jE_\pm \sin\psi) f_1(\theta') - (A_\pm \cos\psi - jC_\pm) f_2(\theta') \quad (3.93)$$

and

$$(I_{\alpha_N})_{M,S} = -(C_\pm + jA_\pm \cos\psi) f_1(\theta') - (D_\pm \cos\psi + E_\pm \sin\psi + jB_\pm) f_2(\theta') \quad (3.94)$$

and

$$\psi = \theta - \theta' \quad (3.95)$$

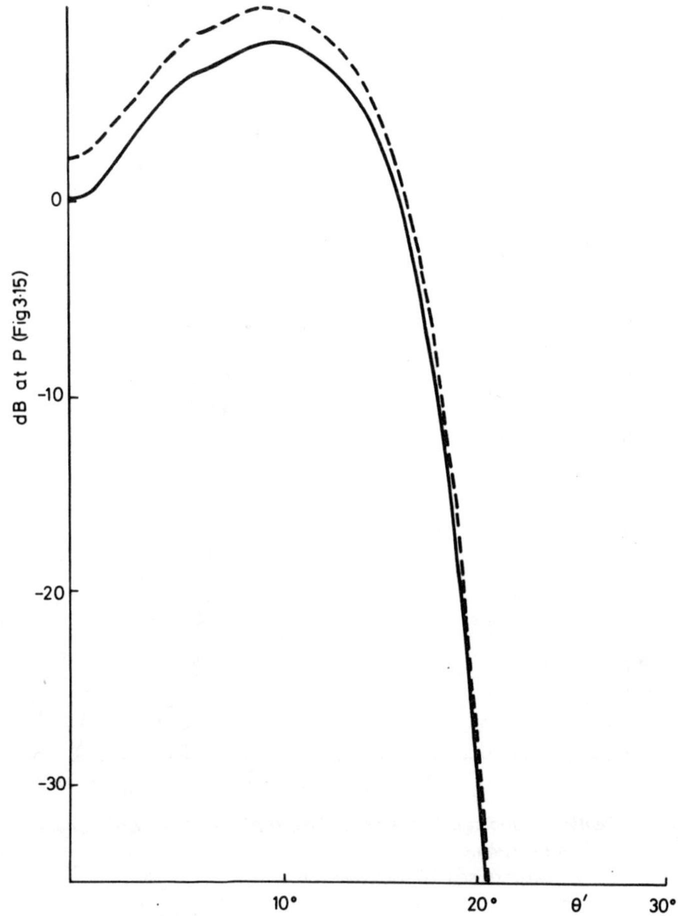

Fig. 3.16. *Realisable and synthesised patterns, E-plane (shown displaced).*
——— realisable
- - - - synthesised 10·4λ reflector

A value of n_L of 35 is deduced from the empirical rule of eqn. 88, and is compatible with the spectrum-generating capabilities of a 10·4λ parabolic reflector. Summation of the Laurent series based on eqns. 92–95 up to this index yields the 'realisable patterns' of Fig. 3.16 (*E*-plane) and Fig. 3.17 (*H*-plane).

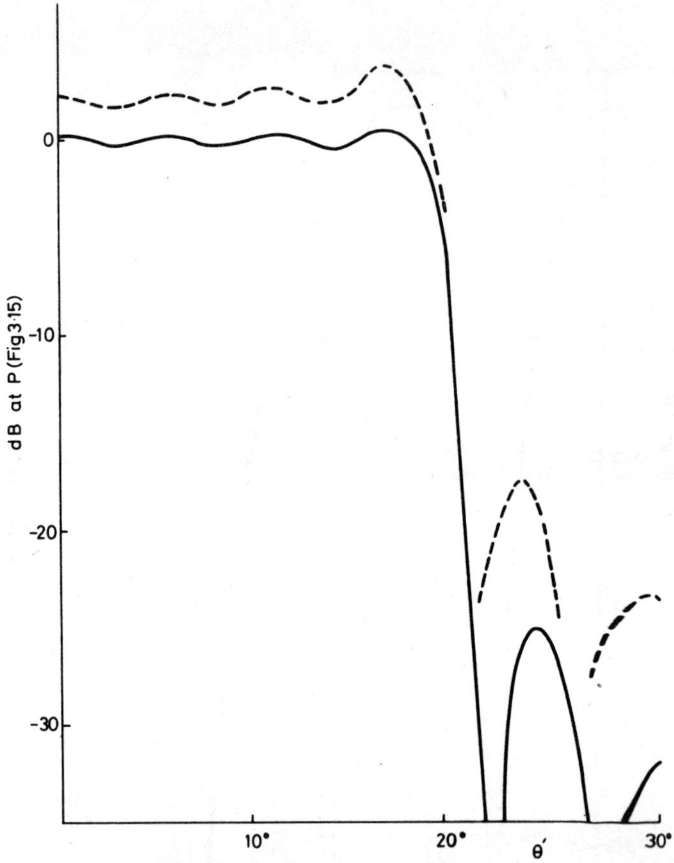

Fig. 3.17. *Realisable and synthesised patterns, H-plane (shown displaced)*
——— realisable
- - - - synthesised 10·4λ reflector

The same spectrum coefficients are now used to evaluate the Uniform expansion of eqns. 80–83 at the paraboloid surface. In this case, however, continuation of the series to a radial index of 35 is only

3.4.5 The general synthesis problem

In the example discussed in the previous Section, the phase of the radiation pattern was constrained to approach a specified curve. However, in many shaped beam antenna applications, this is neither necessary nor desirable. Rather, it is required to generate the best possible approximation to a given *amplitude* characteristic, regardless of phase.

Fortunately, Uniform expansion synthesis can fairly readily be adapted to this role. Thus, initially an arbitrary (but preferably physically reasonable) phase distribution is specified. The numerical efficiency of the expansion allows this initial guess to be improved on by iteration. At each stage the actual radiated pattern associated with the synthesised current is calculated, and the phase of this pattern (together with the *original* ideal amplitude pattern) used as a starting point for the next iteration.

However, in many applications further problems will be encountered. Thus, in the hypothetical example of Section 3.4.4.3, it was a straightforward matter to calculate a current distribution (e.g. Figs. 3.18 and 19) which would generate a radiation pattern with the desired characteristics. However, in practice it would be necessary to limit the current distribution to one of a class that could be realised with one of the more common types of feed horn. The latter restriction implies a roughly parabolic amplitude distribution for the reflector induced currents, unless the horn is multimoded and its aperture size is sufficient that more than one lobe of its radiation pattern is captured by the reflector. Moreover, the phase of the currents induced by a realisable feed must always vary in a fashion which is roughly compatible with the path length from the reflector surface to a 'best-fit' feed phase centre.

The latter constraint may be removed by allowing the reflector profile itself to be a variable in the synthesis. Virtually any type of phase distribution can then be realised. One example in which this type of approach has been adopted is the MARECS antenna.[8, 9] This antenna generates a shaped beam (see Fig. 3.20) suitable for efficient earth coverage from a geostationary satellite. In addition to the amplitude characteristic of the feed horn (an 'ICR' choked circular waveguide, Section 5.1.4), there are other special constraints[8, 10] relating to the particular properties of the shaped beam and the requirement for a matched reflecting surface. No universal or completely methodical approach to a design problem of this complexity has yet been established. In the MARECS case, the optimum solution has been arrived at largely by trial and error techniques which are peculiar to that application. Some final minor adjustments to the position of the feed horn were carried out empirically. These were needed because of the limitations of theoretical modelling for the MARECS antenna. Specifically, the limitations relate to the small size of the reflector (10λ diameter), and the multiple scattering effects inevitable with a central feed horn having an associated support structure adequate to survive a space launch. Multiple scattering effects could not be included in the model, and are thought to represent the major cause of the discrepancy of up to 0·5 dB which exists between the theoretical and measured shaped beam pattern.[9]

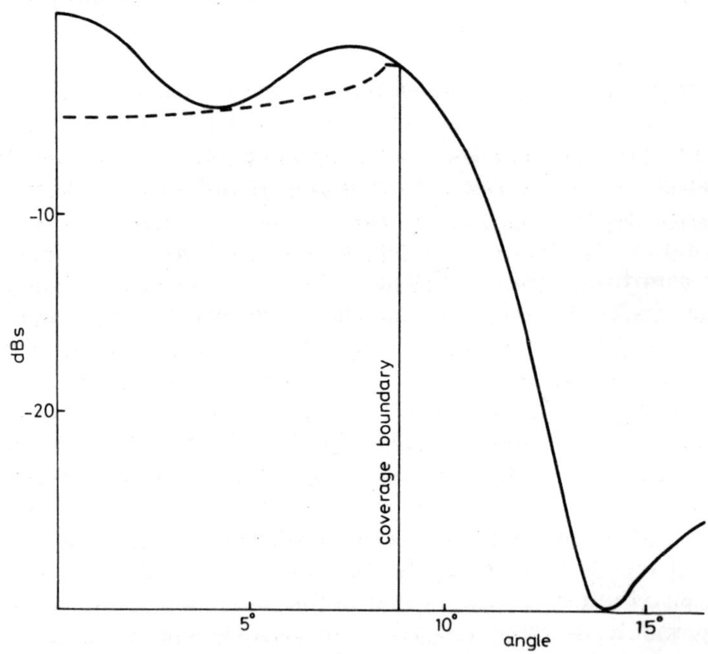

Fig. 3.20. *Radiation pattern synthesised for Earth coverage from a geostationary satellite*

——— computed pattern for shaped reflector antenna

– – – – ideal requirement

Another specialised type of synthesis involving reflector profiles is the high efficiency Cassegrainian feed application, to be discussed in Chapter 7.

3.5 Further properties of Laurent, Taylor and Uniform expansions

3.5.1 Equivalence to KHS vector diffraction

The Laurent series of eqns. 28, 29 and 38 and the Taylor series of eqns. 66, 67 and 68 are direct analytical equivalents of the KHS vector diffraction integrals of Chapter 2, eqns. 2.1 and 2.2. Fundamentally, both types of solution result from invoking Green's vector identity in conjunction with Maxwell's equations. The equivalence has been tested numerically for a number of different problem geometries, encompassing both near and far field calculations.

Some examples of this type of test have been given in References 2 and 11. An agreement of S.W.E. and KHS results to the order of one part in 10^5 has been demonstrated for all vector components of E- and H-fields.

3.5.2 Choice of expansion and co-ordinate origin

3.5.2.1 Use of Laurent, Taylor, or Uniform expansion: In a problem which involves the calculation of the scattered fields of a reflector from its induced currents, it may be possible to use either a Laurent or a Taylor expansion. However, the region of validity is different in the two cases. That is, in general the observation point must lie outside the Laurent minimum sphere of Section 3.2.1, and inside the Taylor maximum sphere of Section 3.3.1. In practice, this means that with an expedient choice of S.W.E. origin the observation point may be in closer proximity to the source for the Taylor case. Neither the Taylor nor the Laurent series is valid actually at the surface of a current sheet source; the Uniform expansion can be used in this special case.

The maximum radial index n_L for the Laurent series relates to the maximum radius of the *source* surface relative to the S.W.E. origin, through eqn. 44 and Fig. 3.4. For the Taylor series, however, the maximum radius of the *observation* surface is the critical parameter. Thus for example in evaluating focal plane patterns (Section 3.3.4), the observation surface will ordinarily be of smaller radius than the source, and the Taylor expansion is more efficient even if the geometry would allow a Laurent expansion. However, for evaluation of scattered patterns in the far field or at great range, the observation surface will be of large radius, favouring the Laurent expansion. Also, only the Laurent expansion can yield an 'all-round' pattern for an antenna.

3.5.2.2 Co-ordinate origin for spherical wave expansion: For problems involving rotationally symmetric reflectors, the number of Laurent spectrum-weighting coefficients will ordinarily be minimised by selecting an origin at the centre of 'the aperture'. The latter in this context can be defined as a plane through the rim of the reflector. However, locating the S.W.E. origin at the reflector vertex enables maximum advantage to be taken of the 'radial index rings' of Fig. 3.5 when calculating the contribution of the centre portion of the reflector to the integral of eqn. 38. Usually this latter choice is slightly more efficient overall in computational terms.

For a Taylor-type expansion, the origin is most advantageously positioned on the axis of rotational symmetry and in the plane containing the periphery of the observation surface. If there is an optical focus, it is beneficial to locate the origin here. This then promotes 'coherent excitation' properties (Section 3.3.3).

For problems which lack circular symmetry, the criteria dictating the optimum position for the origin are less clear-cut. Thus for the offset reflector of Fig. 3.10, a choice of origin on the axis of revolution of the reflector generating surface will result in a separable 'ϕ' integration for eqn. 38 or eqn. 68. For the more usual offset geometries, this is the overriding consideration. However, for Cassegrainian beam waveguide feeds (Chapter 5) the geometry is usually such that the reflector surface is displaced many aperture diameters from the axis of revolution. An axis of revolution origin will then result in slow convergence for the azimuthal expansion. For example $m_L = 80$ for this origin is typical for a beam waveguide application. Under these circumstances it can be more advantageous to locate the origin instead at the centre of the projected aperture of the reflector. This will markedly improve the azimuthal convergence, resulting perhaps in $m_L \approx 5$.

These considerations are summarised in Table 3.6.

Table 3.6 *Criteria dictating choice of S.W.E. co-ordinate centre*

	At centre of projected aperture	On axis of revolution of generating surface	At vertex of reflector	At optical focus point
Laurent	√	√	√	
Taylor		√		√

3.5.3 Category (a) and (b) antennas

For one important class of analytical model, 'pseudo-circular symmetry' criteria apply (Section 3.2.7.1), and only $m = 1$ S.W.E. modes need be considered. However, for a practical reflector antenna, even when the principal scatterers are rotationally symmetric, there are usually parts of the structure such as the feed horn support struts which violate the criteria. A more comprehensive azimuthal expansion must then be used. That is, in the most general case, the only limitation on the 'm' index is the all-embracing constraint of eqn. 15; the latter corresponds to a mathematical property of Legendre functions in that P_n^m is only non-zero for $m \leqslant n$.

One exception relates to the boresight point. Only the $m = 0$ and $m = 1$ modes are non-zero for $\theta = 0$. In the far field, only the $m = 1$ modes contribute. Moreover, for small angles off boresight, the azimuthal expansion is always highly convergent, and

$$m_L \ll n_L \tag{3.96}$$

However, even for large angles off boresight there is a wide class of antennas designated as 'category (*a*)' which possess sufficient natural geometrical symmetry to limit the azimuthal expansion to perhaps 10 or 20 modes. This applies even when dimensions are large in comparison to the wavelength, so that n_L takes on values of several hundred. Category (*a*) properties have been firmly established from measured near-field data for several different types of reflector antenna.[7] One of these has a noncircularly symmetric profile deliberately introduced for the purposes of generating an elliptical cross-section beam.[12] The S.W.E. technique is well adapted to antennas of this type, since it takes full advantage of the inherent approximate circular symmetry.

For other ('category (*b*)') applications, the full S.W.E. mode set as defined by eqn. 15 must be used. Category (*b*) problems include those in which the z-axis of the expansion is substantially offset or tilted away from the natural boresight axis of the antenna aperture (see Section 3.5.2). They also include the phased array case where radiation occurs from discrete centres rather than in a continuous fashion.

3.5.4. Use of numerical integration rules

The evaluation of eqn. 38 or similar integrals defining the S.W.E. spectrum is usually carried out numerically using a suitable integration rule.

If the matching surface (or source surface in the case of a source matched formulation) constitutes some part of a surface of revolution, then the $r - \theta$ or 'radial' integration can be separated out (Section 3.2.6). A radial integration grid of some 2 to 2·5 n_L points will then be required for convergence of Simpson's rule.

For the ϕ or 'azimuthal' integration, the integrand may be 'band limited' in the sense of the theory of Fourier analysis. This will occur whenever the ϕ integration path is a closed annulus on the reflector surface. It is then expedient to use the trapezoidal rule, this being the most economical in terms of the number of azimuthal grid points required. Also, for a given θ, the ϕ integral amounts to a Fourier transform. Thus, the 'fast Fourier transform' technique can be used if desired, in place of numerical integration. However, often the number of azimuthal modes m_L is insufficient to fully justify the use of the FFT.

If the periphery of the matching surface is not circular and coaxial with the axis of revolution, the integration paths may be open curves. An illustration of the open curve type of problem has been given in Section 3.3.4. In that example (focal plane pattern for offset reflector), numerical integration proved unnecessary. In this particular case, this was because the integrand was of a sufficiently simple form to permit an analytical evaluation. However in general (and if an off-axis wave incident on the offset reflector were considered), analytical evaluation would not be feasible. Furthermore, in an open curve situation the integrand is no longer band-limited. Under these circumstances, the trapezoidal rule is inappropriate and a higher order quadrature rule should be used. These considerations are summarised in Table 3.7.

Table 3.7 *Integration rules for calculating S.W.E. coefficients*

Problem	Radial (r, θ) integration	Azimuthal (ϕ) integration
Circular symmetry case	Simpson	
Asymmetric illumination law but $\theta =$ constant at scatterer periphery	Simpson	Trapezium ($\equiv$ FFT)
Asymmetric periphery	Simpson	Simpson

3.5.5 Comparison with alternative vector diffraction techniques

By way of conclusion to the presentation of the concepts of antenna diffraction theory as outlined in Chapter 2 and the present chapter, a brief 'over-view' of the three principal analytical techniques, KHS, GTD and S.W.E., will be given.

As will have emerged in this chapter, an important attribute of the S.W.E. is its scope and versatility. That is, a unified S.W.E. analytical method accommodates not only applications involving theoretical modelling of scattering structures, but also permits the introduction of experimental data from test range measurements, and caters for inverse problems where the aim is to synthesise a desired radiation pattern.

Included under the category of theoretical modelling is the calculation of scattered patterns of reflecting surfaces, and the calculation of feed horn patterns from a theoretical aperture distribution. The latter topic will be treated in Chapter 5. S.W.E. radiation patterns can equally readily be calculated at near or far field range, irrespective of the presence of caustics or an optical focus. The only constraints are the minimum and maximum spheres: for practical applications, these rarely amount to a significant limitation on the method. An example of the 'theoretical modelling' type of problem is the transmitting and receiving analysis of Cassegrain antennas, to be discussed in Chapter 7.

Where the starting point is experimental data, the expansion can be used to rigorously characterise the antenna being tested, or to enable fields to be predicted at an alternative range. For example, in the case of measured data for a feed horn, the S.W.E. enables the horn fields to be predicted at a radius corresponding to the surface of the reflector it is to illuminate (see Chapter 5). Another example is the characterisation of beam waveguide Cassegrainian feeds from experimental data (see Chapter 5).

References 13–16 and 7 describe a spherical wave test facility which employs on-line digital near field measurements. This facility has been constructed for testing shaped beam satellite antennas.[9] It uses the S.W.E. both to effect smoothing of random errors and to calculate the far-field pattern from near-field data. This results in a superior overall measurement accuracy in comparison with conventional testing techniques. *A* ± 0·07 dB accuracy has been verified in a particular instance.[16]

Included under the final category of inverse and synthesis problems are those for which it is required to establish the degree of approximation to a desired pattern that is achievable with a radiating source of given aperture diameter. In other applications it may further be required to determine the current distribution residing on a specified curve surface which will launch this pattern. One special type of synthesis procedure for Cassegrain antennas forms the subject of Chapter 7.

Although the three categories of problem under discussion are apparently quite different, actually they are remarkably similar at the level of an S.W.E. computer algorithm. For example, the core of the computer software for the offset reflector focal plane calculations of Section 3.3.4 is virtually the same as that for the hyperboloid of Section 3.2.6, and for the near-field far-field transformation of measured field data of Reference 7. The only fundamental difference is associated with the type of radial function used. This in turn relates to the starting values for the recursive relationships of eqns. 16 and 17.

By way of contrast, only the first two applications are amenable to a KHS vector diffraction treatment, whereas only the second could feasibly be tackled by GTD.[17] In the GTD case, a minor alteration in problem geometry or reflector profile might well require a complete reformulation.

With any computer implementation of vector diffraction theory, an important feature is 'software validation', e.g. checking out program modules to ensure there are no coding errors. This is particularly essential in applications where 'second order' parts of the solution (e.g. the crosspolar pattern) are of importance. Because the S.W.E. is an exact formulation of electromagnetism, it remains valid for limiting case geometries. For example, most problems can be reduced in the limit to the 'scattering from an infinite plane conductor' test proposed by Ludwig.[3] Moreover, any one S.W.E. formulation tends to differ from others only in respect of a small portion of the coding, and the validation procedure can be restricted to this portion.

An expansion technique, however, can be cumbersome when dealing with a particularly simple problem. Such simplifications of the general S.W.E. that are possible rarely suffice to bring the method within the scope of a slide rule or programmable calculator. The S.W.E. technique is better suited to all-purpose programmes for a digital computer.

In these latter respects the S.W.E. is of course the antithesis of the GTD method. The philosophy of the GTD is to attempt to find closed form representations of specific vector diffraction problems by introducing certain asymptotic approximations. The latter frequently tend to limit the generality of the formulation.

To summarise these conclusions, in Table 3.8 an 'order of merit' has

Table 3.8 *Comparison of alternative numerical techniques for reflector antenna applications*

	Feature	SWE/ PO	KHS/ PO	GTD
1	Scope: range of problems for which technique usable	1	2	3
2	Feasibility of universal software modules	1	2	4
3	Feasibility of 'slide rule' calculations	4	2	1
4	Interfacing with measured data	1	3	4
5	Computer CPU time	2	3	1
6	Low edge illuminations	1	1	4
7	Traceability of approximations	1	1	2
8	Ease of checkout using limiting cases	1	2	3
9	Inverse and synthesis problems	1	2	3
10	Useful simplified forms	4	1	1

been assigned to each of the three analytical techniques in respect of criteria as listed. An order of merit of '1' indicates the most preferable technique in relation to the particular criterion applicable.

References

1 STRATTON, J. A.: 'Electromagnetic theory' (McGraw-Hill, 1941)
2 WOOD, P. J.: 'Spherical waves in antenna problems', *Marconi Rev.*, 1971, **34**, pp. 149–172
3 LUDWIG, A. C.: 'Calculation of scattered patterns for asymmetrical reflectors'. J.P.L. Technical Report 32–1430, Feb. 15th, 1970
4 WOOD, P. J.: 'Comment on numerical check on the accuracy of spherical wave expansions', *Electron. Lett.*, 1972, **8**, pp. 237–238
5 HARRINGTON, R. F., and MAUTZ, J. R.: 'Theory of characteristic modes for conducting bodies', *IEEE Trans.*, 1971, **AP-19**, pp. 622–628
6 McLACHLAN, R.: 'Complex variable and operational calculus with technical applications' (Cambridge University Press, 1939)
7 WOOD, P. J.: 'The prediction of antenna characteristics from spherical near field measurements – Part 1 Theory,' *Marconi Rev.*, 1977, **40**, pp. 42–68
8 WOOD, P. J.: 'Shaped beam reflector design for earth coverage applications'. IEE Antennas for Aircraft and Spacecraft Conference, London, 1975, pp. 47–52
9 COWAN, J. H.: 'A shaped beam antenna for a maritime communications satellite'. *ibid.*, pp. 101–106
10 WOOD, P. J.: 'Optimisation of antenna gain for earth coverage from a geostationary satellite', *Electron. Lett.*, **10**, pp. 227–228
11 WOOD, P. J.: 'Spherical harmonic expansions in near field aerial problems', *ibid.*, 1970, **6**, pp. 535–536
12 DORO, G., and SAITO, A.: 'Dual Polarisation Antennas for OTS'. IEE Antennas for Aircraft and Spacecraft Conference, London, 1975 pp. 76–82b
13 WOOD, P. J., and LOCKETT, N. J. L.: 'A new type of test range for polar diagram measurements', *ibid.*, pp. 186–191
14 WOOD, P. J.: 'Principles of antenna measurements using on-line computer processing with spherical probing'. Conference on Antenna Testing Techniques, Holland, June 1977, ESA publication SP 127, pp. 3–8
15 WOOD, P. J.: The MAROTS antenna measurement facility'. *ibid.*, pp. 69–76
16 WOOD, P. J.: 'The prediction of antenna characteristics from spherical near field measurements; PART II, experimental validation', *Marconi Rev.*, 1977, **40**, pp. 117–155.
17 SORENSEN, O., and RUSCH, W. V. T.: 'Application of the geometrical theory of diffraction to Cassegrain subreflectors with laterally defocussed feeds', *IEEE Trans.*, 1975, **AP-23**, pp. 698–702

Chapter 4

Field correlation methods

4.1 Robieux's theorem

4.1.1 Statement of theorem

According to Robieux's theorem,[1] the coupling between two antennas 1 and 2 (Fig. 4.1) may be specified by the complex transmission coefficient T,

$$T = \frac{1}{2} \int_S E_1 \times H_2 - E_2 \times H_1 \cdot dS \tag{4.1}$$

The symbol T here denotes a voltage ratio, although Robieux's original paper defines T as a complex power ratio (T^2 in the present notation). S can be any closed surface which completely encloses one antenna whilst excluding the other.

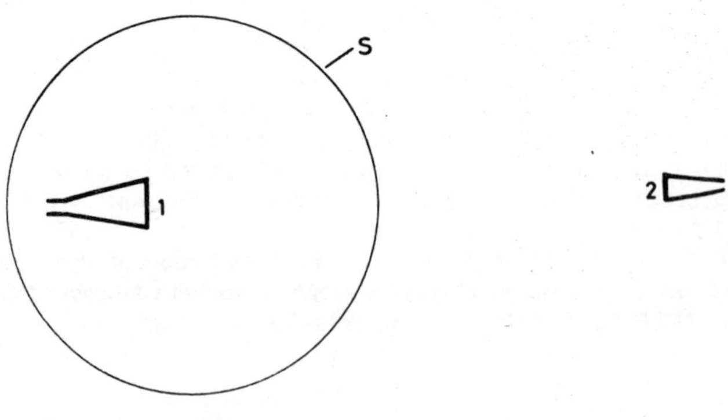

Fig. 4.1. *Robieux's theorem*

Robieux's theorem is a very general, all-embracing relationship from which many of the radiation formulas of Chapter 2 can be derived.[1]

For example, eqn. 2.5 for the far fields radiated by an aperture can readily be deduced by taking S as the aperture surface, $\boldsymbol{E}_1 : \boldsymbol{H}_1$ as the aperture field, and $\boldsymbol{E}_2 : \boldsymbol{H}_2$ as a plane wave field launched by a distant probe.

4.1.2 Conditions for validity

The conditions associated with Robieux's theorem are:

(*a*) Each antenna must have, in concept at least, recognisable terminals connected by some form of single mode transmission line (Fig. 4.1). The transmission coefficient T is then a property of a two-port microwave network, relating the forward propagating waves in the excited and load transmission lines.
(*b*) The field $\boldsymbol{E}_1 : \boldsymbol{H}_1$ is defined as the radiated field for unit power input to terminals 1, *in the presence of the (passive) antenna 2.* A similar definition applies to $\boldsymbol{E}_2 : \boldsymbol{H}_2$.
(*c*) The theorem applies to two antennas in free space; it does not allow for any additional scattering object.

These restrictions can only be relaxed when the separation between antennas is very large, so that interaction effects are negligible.

4.1.3 Robieux's theorem and reflector scattering

In anticipation of possible extensions of the method to single and multiple reflector antennas, it is natural to attempt to transform eqn. 1 so that S surrounds one of the reflectors. Ultimately, it would then be possible to 'collapse' S to the reflector surface, in a manner analogous to the derivation described in Section 3.2.1.

Fig. 4.2 depicts a paraboloid antenna comprising a horn A, and a reflector B. To apply Robieux's theorem, 'terminals' are explicitly assigned to the far-field pattern by introducing the remote probe C. For the purposes of the theorem, there are now two antennas with terminals near A and C, and a reflector which must be regarded as a part of one or other of them. Take

$$\text{antenna } 1 \equiv \text{A}$$

$$\text{antenna } 2 \equiv \text{C} + \text{B}$$

with S enclosing antenna 2. The surface integrals over *ab* and *cd* of Fig. 4.2 must cancel; hence the integration surface for eqn. 1 reduces to the two surfaces S_1 and S_2.

At S_2

$$\boldsymbol{n} \times \boldsymbol{E}_2 = 0$$

(C + B launches a field obeying perfect conductor boundary conditions at the reflector) and

$$n \times E_1 = 0$$

(field launched by A *in the presence of antenna 2* obeys perfect conductor boundary conditions at the reflector).

Thus, only the surface integral over S_1 can be finite. That is, attempting to locate the integration surface at the reflector has led to an essentially trivial result: namely, the coupling into the probe is proportional to the level of the paraboloid far field pattern times the probe effective receiving area. Although only one particular instance has been described, this type of limitation applies quite generally for Robieux's theorem.

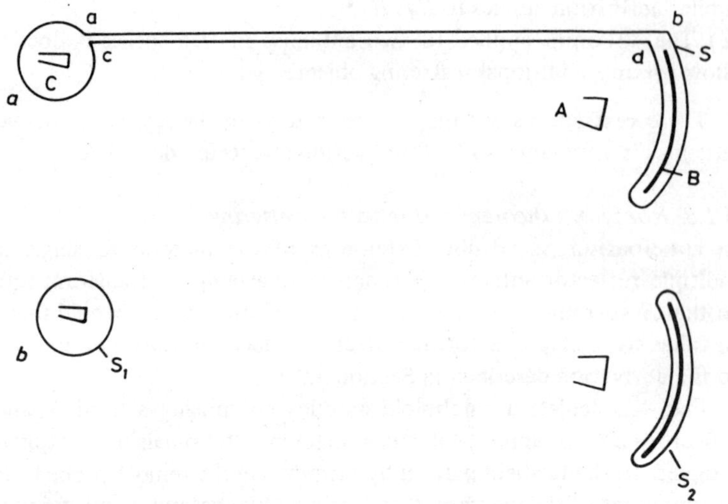

Fig. 4.2. *Reduction of Robieux's theorem for a reflector antenna*

4.2 Field correlation theorem

4.2.1 Statement of theorem

According to the field correlation theorem,[2] the voltage efficiency of an antenna is specified by

$$\eta = \frac{\iint_S (E_r \times H_t - E_t \times H_r) \cdot dS}{2[R_e \int_{S_\infty} E_t \times \overset{*}{H}_t \cdot dS]^{1/2} [R_e \int_A E_r \times \overset{*}{H}_r \cdot dS]^{1/2}} \tag{4.2}$$

Here dS has the sense of the outward normal to the closed surface S, and η is measured relative to an ideal uniformly illuminated aperture of area A. That is, the power gain of the antenna may be represented as

$$G = |\eta|^2 \frac{4\pi A}{\lambda^2} \tag{4.3}$$

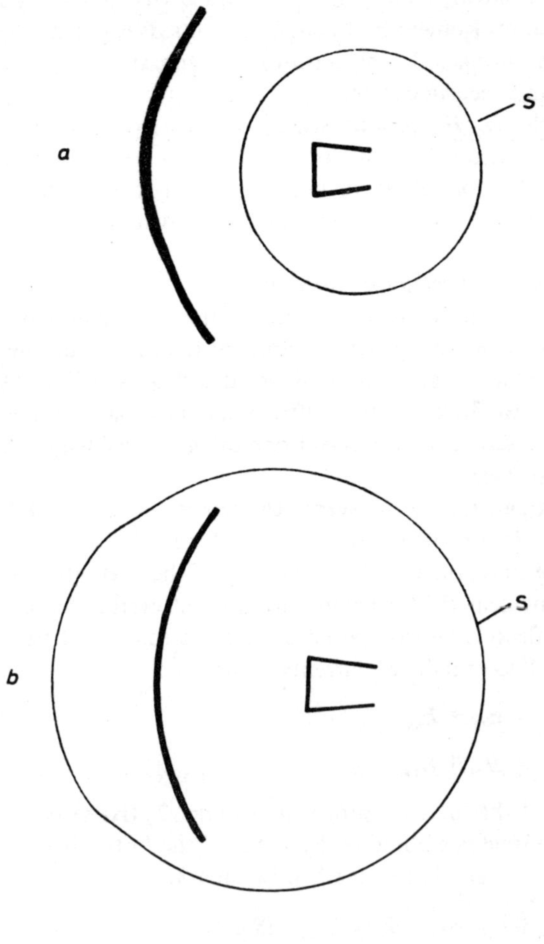

Fig. 4.3. *Correlation surfaces for a front-fed antenna*

Algebraically, eqn. 2 is very similar to Robieux's theorem. However, the conditions for its validity are as follows:

(*a*) The receiving field $E_r : H_r$ is to be interpreted as including scattering/interaction contributions from all parts of the antenna. However, the transmitting field $E_t : H_t$ by definition includes only the radiated field of sources internal to S, e.g. the feed horn of Fig. 4.3*a* or the feed/reflector antenna of Fig. 4.3*b*. Interaction effects associated with sources external to *S are not to be taken into account.*
(*b*) The surface *S* must lie in free space and enclose terminals associated with the transmitting field (Fig. 4.3). There is no alternative choice for *S* in the sense of Robieux's theorem. The coefficient η relates to the forward wave propagating power into the terminals, when the antenna is operated in a receiving state.
(*c*) The field $E_r : H_r$ is intrinsically independent of terminals. This implies that an explicit definition will need to be introduced to permit this field to be normalised to unit power. In effect, the definition is provided by the second integral of the denominator of eqn. 2.

4.2.2 *Treatment of reflector scattering*

It is usually advantageous to arrange that the field correlation surface *S* is coincident with some physical feature of the antenna. One possibility is the final radiating aperture; however this choice will lead to a similar formulation to KHS vector diffraction. For a multiple reflector antenna, an alternative is to select one of the subsidiary reflectors as a correlation surface.

For example, for the Cassegrainian geometry of Fig. 4.4, we apply the field correlation theorem with $S = S_1$. When the antenna is operated in a receiving state, the field in the vicinity of the subreflector comprises in part an incident field generated by the scattering of a plane wave by the main reflector. Superimposed on this is a scattered field generated by the subreflector induced currents. Thus:

$$E_r = E_{ri} + E_{rs}$$

and

$$H_r = H_{ri} + H_{rs}$$

Because of the linearity properties of eqn. 2, these two constituent fields can be treated separately. By virtue of the field theorem discussed in Section 3.2.1. and Reference 2 of Chapter 3.

$$\int_{S_2} \{E_{rs} \times H_t - E_t \times H_{rs}\} \cdot dS = 0$$

This result enables the integration surface for eqn. 2 to be transformed for the scattered constituent field, so that in respect of this field the numerator of the equation becomes

$$N = \int_{S_3} \{E_{rs} \times H_t - E_t \times H_{rs}\} \cdot dS \tag{4.4}$$

Eqn. 4 would be identically zero if all the integrand fields were continuous across the subreflector, by virtue of the nature of S_3 (Fig. 4.4). Because the field correlation theorem has been invoked as opposed to Robieux's theorem, $\boldsymbol{E}_t : \boldsymbol{H}_t$ may be taken as being continuous. This field constitutes the feed horn radiation unperturbed by the presence of the subreflector. However, according to physical optics, $\boldsymbol{H}_{rs}$ is discontinuous, e.g.

$$D_{+-}(\boldsymbol{n} \times \boldsymbol{H}_{rs}) = 2\boldsymbol{n} \times \boldsymbol{H}_{ri}$$

so that

$$\eta_s = \frac{\int_{S_{sub}} (\boldsymbol{H}_{ri} \times \boldsymbol{E}_t) \cdot d\boldsymbol{S}}{[R_e \int_{S_\infty} \boldsymbol{E}_t \times \overset{*}{\boldsymbol{H}}_t \cdot d\boldsymbol{S}]^{1/2} [R_e \int_A \boldsymbol{E}_r \times \overset{*}{\boldsymbol{H}}_r \cdot d\boldsymbol{S}]^{1/2}} \tag{4.5}$$

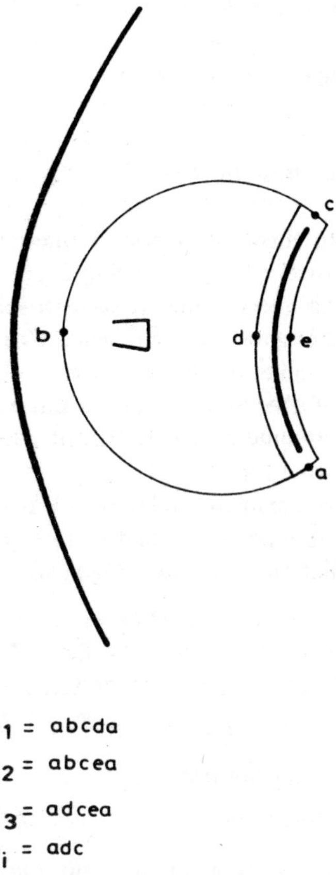

Fig. 4.4. *Correlation surfaces for dual reflector antenna*

Here the '*S*' suffix for η indicates that the antenna performance is specified in respect of a mechanism of coupling between the main reflector and the feed horn involving only subreflector scattering. The small perturbation associated with operation of feed and main reflector with subreflector removed (see Section 2.2.7) is not taken into account.

Unlike eqns. 1 or 2, eqn. 5 is asymmetric with respect to *E* and *H*. This relates to the 'electric wall' properties assumed for the subreflector. An alternative formulation with

$$\int_{S_{sub}} \boldsymbol{E}_{ri} \times \boldsymbol{H}_t \cdot d\boldsymbol{S}$$

would apply to a magnetic wall scatterer. In the special case of the application of eqn. 5 to a single reflector antenna, the distinction is ordinarily unimportant. That is, electric and magnetic versions differ only to the extent that the PO transmitting and receiving solutions are nonreciprocal. For application to an intermediate reflector however, the formulation of eqn. 5 should always be used.

4.3 Slope correlation theorem

The field correlation theorem of eqn. 5 may be used to calculate a complete far-field pattern by introducing a series of off-axis incident waves. However, it is most efficient in applications where the only requirement is to calculate boresight gain. For a Cassegrain antenna with rotationally symmetric reflectors, the diffraction calculations required are then of the '$m = 1$' type discussed in Section 3.2.7.1, irrespective of the symmetry properties of the specified feed horn pattern.

A complementary parameter relevant to monopulse or autotrack antennas is boresight slope. We obtain from eqn. 5 after applying the theorem of differentiation under the integral sign

$$\left[\frac{\partial \eta_s}{\partial \Theta}\right]_{\Theta=0} = \frac{\int_{S_{sub}} \left[\dfrac{\partial \boldsymbol{H}_{ri}}{\partial \Theta}\right]_{\Theta=0} \times \boldsymbol{E}_t \cdot d\boldsymbol{S}}{[R_e \int_{S_\infty} \boldsymbol{E}_t \times \overset{*}{\boldsymbol{H}}_t \cdot d\boldsymbol{S}]^{1/2} \, [R_e \int_A \boldsymbol{E}_r \times \overset{*}{\boldsymbol{H}}_r \cdot d\boldsymbol{S}]^{1/2}} \tag{4.6}$$

At a point specified by the vector

$$\mathbf{p} = x\boldsymbol{a}_x + y\boldsymbol{a}_y + z\boldsymbol{a}_z$$

on the main reflector surface, an incident plane wave may be represented as

$$(E_r)_{main} = \boldsymbol{a}_R \times \boldsymbol{a}_\psi \exp(jk\boldsymbol{\rho}\cdot\boldsymbol{a}_R)$$

$$(H_r)_{main} = \boldsymbol{a}_\psi \exp(jk\boldsymbol{\rho}\cdot\boldsymbol{a}_R)$$

Here the direction of propagation of the plane wave is specified by the unit vector $\boldsymbol{a}_R$, and its polarisation by the unit vector $\boldsymbol{a}_\psi$,

$$\boldsymbol{a}_\psi = \boldsymbol{a}_\Theta \cos(\Phi - \Phi_0) - \boldsymbol{a}_\Phi \sin(\Phi - \Phi_0) \tag{4.7}$$

Thus

$$\left[\frac{\partial H_r}{\partial \Theta}\right]_{main,\,\Theta=0} =$$

$$\{jk(x\cos\Phi + y\sin\Phi)\boldsymbol{a}_\psi - \cos(\Phi - \Phi_0)\boldsymbol{a}_z\}\exp(jkz) \tag{4.8}$$

specifies a 'slope wave' which provides the starting point for a vector diffraction calculation to determine $\partial H_r/\partial\Theta$ at the surface of the sub-reflector.

4.4 Applications

One application in which either Robieux's theorem (eqn. 1) or the field correlation theorem (eqn. 2) is equally appropriate concerns the optimisation of the geometry of feedhorn structures.[3, 4] For maximum power transfer, the aperture fields of the horn should ideally be conjugate matched to the focal plane pattern corresponding to an on-axis incident wave.

The focal plane approach can also be extended to the design and analysis of autotrack systems of the peripheral horn type. This subject is treated in Chaper 8.

The reflector field correlation theorem of eqn. 5 is used in the synthesis of reflector profiles for shaped beam Cassegrain antennas as described in Chapter 7. The calculation of reflector V.S.W.R. (Chapter 9) is another application. Finally, the method has been used to estimate phase error correction factors for insertion loss of a Cassegrainian beam waveguide feed, when using short-circuit plate measurements.[5]

References

1 ROBIEUX, J.: 'Lois Generales de la Liason entre Radiateurs d'Ondes', *Ann. Radioelectr.*, 1959, **14**, pp. 187–229

2 WOOD, P. J.: 'Field correlation theorem with application to reflector aerial diffraction problems', *Electron. Lett.*, 1970, **6**, pp. 326–327
3 VU, T. B.: 'Optimisation of efficiency of reflector antennas: approximate method', *Proc. IEE.*, 1970, **117**, pp. 30–34
4 POULTON G. T., and CLAYDON, B.: 'The determination of crosspolarisation by the use of focal region fields', *Electron. Lett.*. 1973. **9**. pp. 568–569
5 WOOD, P. J.: 'A beam waveguide feed for an 11/14 GHz antenna', IEE International Conference on antennas and propagation, Nov. 1978, pp. 223–229

Chapter 5

Numerical models for reflector antenna feed systems

5.1 Analytical models

5.1.1 Spherical wave expansion for conical feed horn

The first step in analysing a reflector system is ordinarily to establish a numerical or analytical representation of the fields of the primary feed. The importance of this stage of the analysis should not be under-estimated, since simplifications made in modelling the feed horn may sometimes comprise the principal limiting factor specifying the accuracy to which the overall antenna performance can be evaluated.

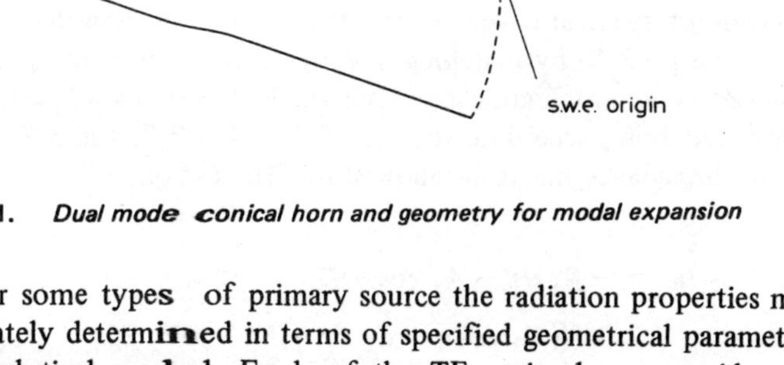

Fig. 5.1. *Dual mode conical horn and geometry for modal expansion*

For some types of primary source the radiation properties may be accurately determined in terms of specified geometrical parameters via an analytical model. Feeds of the TE_{11} circular waveguide excited conical horn family come under this category. These include the smooth wall dominant mode conical horn, the dual mode TE_{11}/TM_{11} horn of Potter[1] and the corrugated conical horn.

For the smooth-walled horn of Fig. 5.1, the transverse fields at P on the spherical aperture cap are

$$\mathrm{TE}_{11}\begin{cases} H_\theta^c = J_1'(P_E x)\sin\phi & (5.1)\\ H_\phi^c = \{J_1(P_E x)/P_E x\}\cos\phi & (5.2)\\ \beta_E E_\theta^c = kH_\phi^c & (5.3)\\ \beta_E E_\phi^c = -kH_\theta^c & (5.4)\end{cases}$$

and

$$\mathrm{TM}_{11}\begin{cases} H_\theta^c = \alpha'\{J_1(P_M x)/P_M x\}\sin\phi & (5.5)\\ H_\phi^c = \alpha' J_1'(P_M x)\cos\phi & (5.6)\\ kE_\theta^c = \beta_M H_\phi^c & (5.7)\\ kE_\phi^c = -\beta_M H_\theta^c & (5.8)\end{cases}$$

Here β_E and β_M are the modal propagation coefficients

$$\beta_M^2 = k^2 - (P_M/a)^2 \tag{5.9}$$

$$\beta_E^2 = k^2 - (P_E/a)^2 \tag{5.10}$$

and

$$x = \psi/\psi_H \tag{5.11}$$

The quantities P_E and P_M are the Bessel function roots 1·841 and 3·832, respectively. The parameter 'a' is the horn aperture radius.

The constant α' has been included in the formulation to cater for an arbitrary combination of TE_{11} and TM_{11} modes. It is related to the moding ratio α as defined by Potter[1] as follows:

$$\alpha' = \frac{J_1(P_E)ka}{J_0(P_M)P_E}\alpha \tag{5.12}$$

The Laurent spectrum weights for the horn can now be derived according to eqn. 3.46 by matching directly to the aperture cap fields. The analysis is almost identical to that of Section 3.4.4.3, with the exception that now, according to eqns. 5.3, 5.4, 5.7, and 5.8 a non-unity wave impedance has to be allowed for. Thus eqn. 3.92 holds, but with

$$(I_{\alpha_N})_E = -B_\pm H_\theta^c - A_\pm \cos\gamma H_\phi^c + j\{(D_\pm\cos\gamma + E_\pm\sin\gamma)E_\phi^c - C_\pm E_\theta^c\} \tag{5.13}$$

and

$$(I_{\alpha_N})_M = -C_\pm H_\theta^c - (D_\pm\cos\gamma + E_\pm\sin\gamma)H_\phi^c + j(B_\pm E_\theta^c - A_\pm\cos\gamma E_\phi^c) \tag{5.14}$$

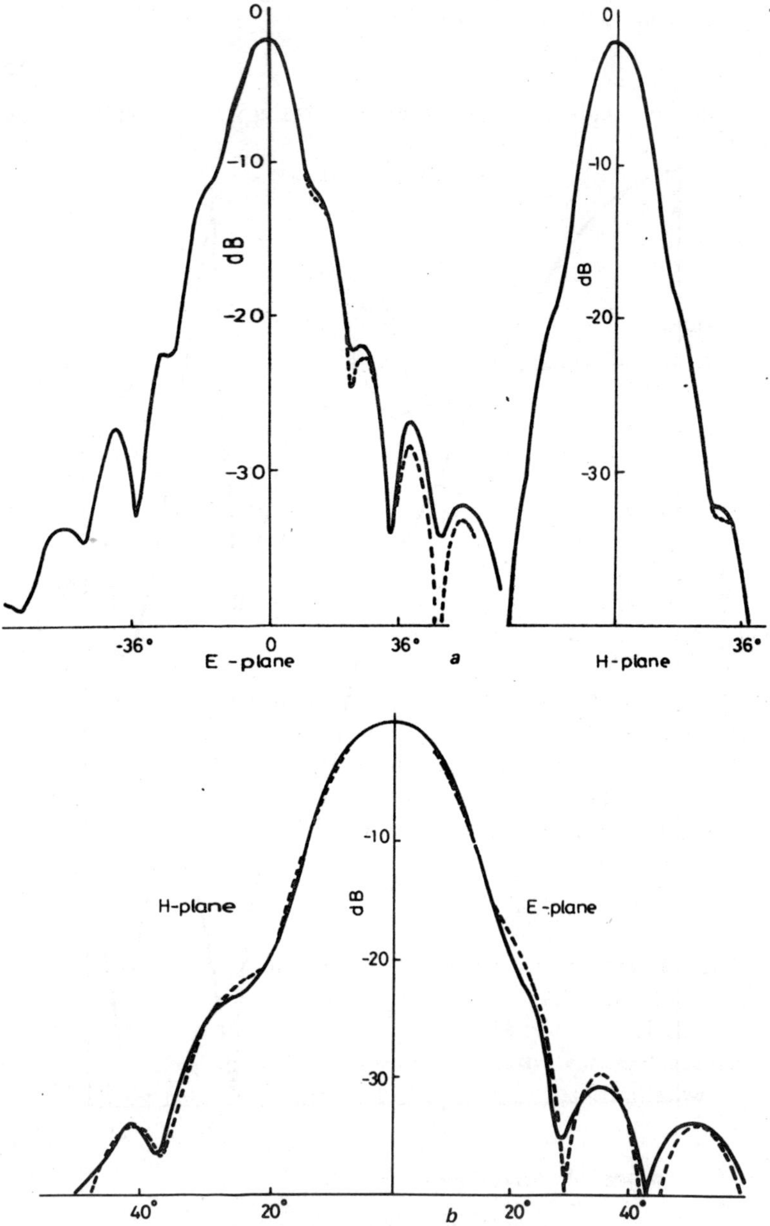

Fig. 5.2. *Radiation patterns of two small conical horns*
(*a*) $a = 2{\cdot}874\lambda$ $\psi_H = 11{\cdot}16°$ TE_{11} conical horn
(*b*) $a = 2{\cdot}4\lambda$ $\psi_H = 9{\cdot}5°$ dual mode horn
——— measured
– – – – predicted

replacing eqns. 3.93, 3.94, where

$$\gamma = \theta + \psi \qquad (5.15)$$

Both the fields to be matched H_θ^c etc. and the spherical wave

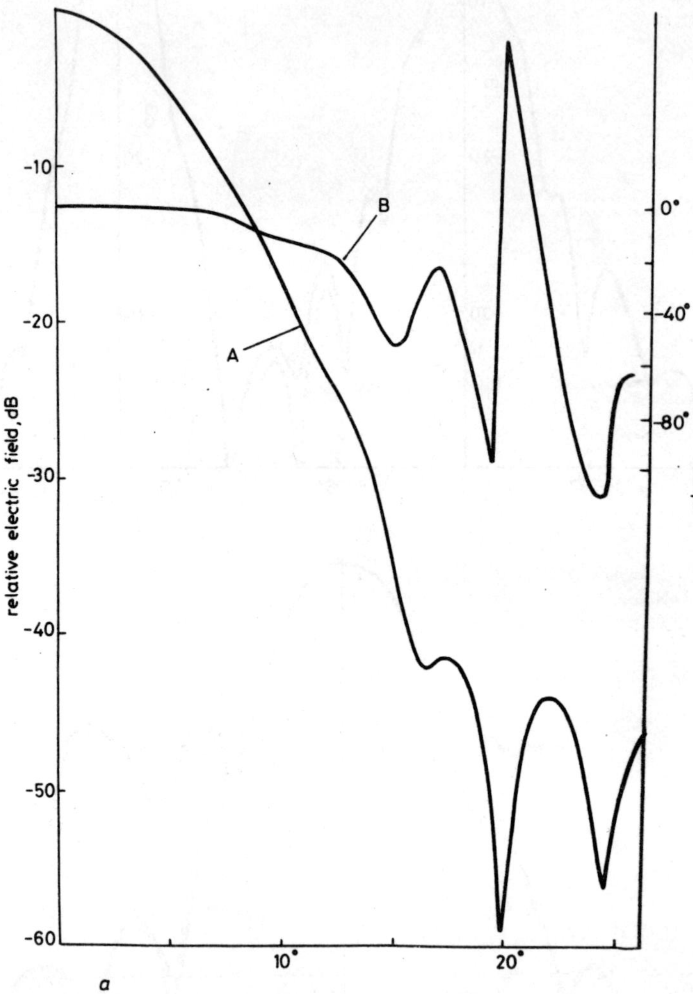

Fig. 5.3. *a* predicted fields, E-plane

functions $A_\pm$ are constant phase quantities. Only real arithmetic is involved in evaluating the spectrum weights. In fact, as an efficient numerical technique for calculating the near fields of a conical horn, the 'spherical cap S.W.E.' would appear to be without rival at present. Its importance, however, lies not so much in its application to dual mode horns, but to the corrugated conical horns commonly used in Cassegrainian reflector systems, as will be described in the following sections.

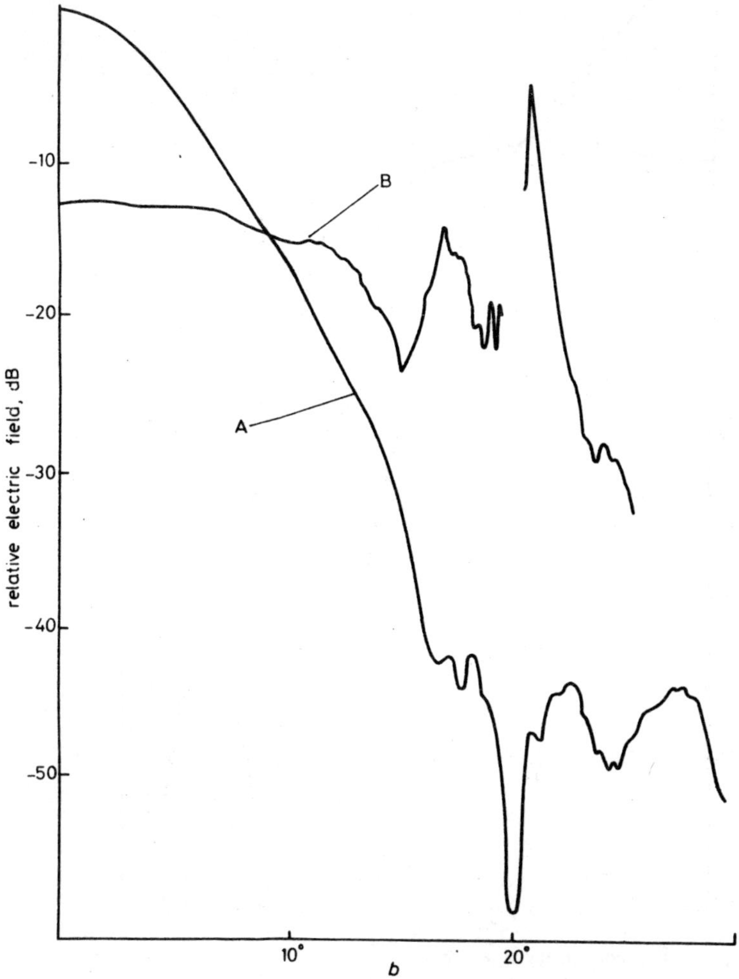

Fig. 5.3. *b* measured fields (Hess & Liesenkotter Ref. 2), E-plane

5.1.2 Applications of conical horn analytical model

Smooth-walled horns

In a real conical horn some excitation of higher order modes and exterior currents will inevitably occur as a result of the discontinuity at the horn aperture. Moreover, the waveguide mode formulations of eqns. 5.1–5.8 are not exact for finite flare angles.

Fig.5.2*a* demonstrates agreement down to −30 dB in the *H*-plane

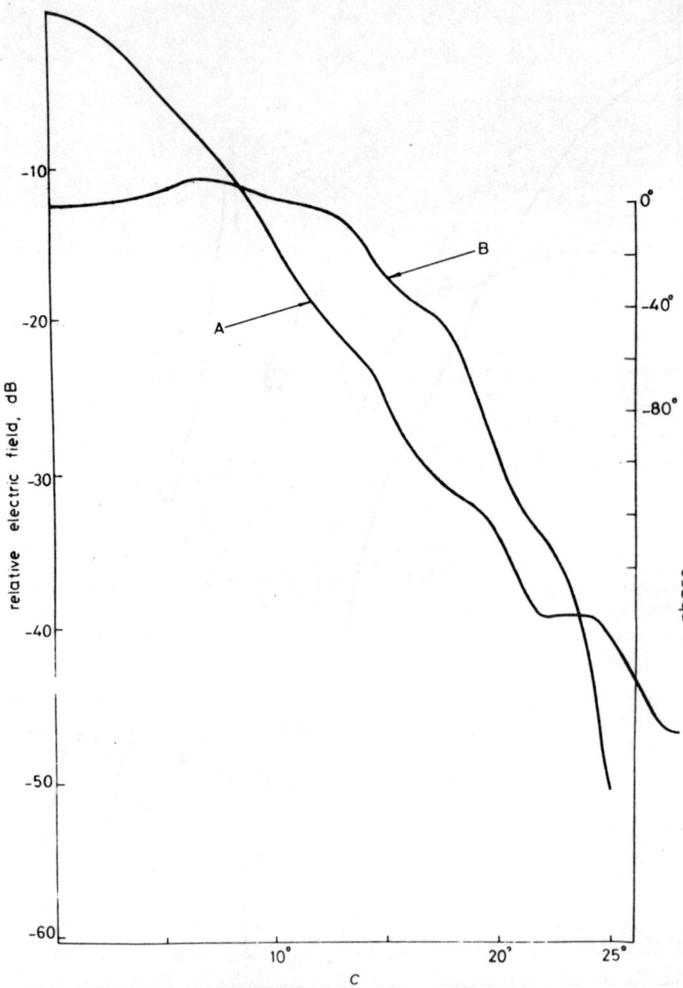

Fig. 5.3. *c* **predicted fields, H-plane**

between the theory of Section 5.1.1 and experimental data for a dominant mode conical horn ($a = 2{\cdot}87\,\lambda$). For this type of horn the aperture rim is illuminated at the highest level in the *E*-plane. Thus exterior currents maximise in this plane, probably causing the discrepancies in the *E*-plane sidelobe region evident from Fig. 5.2*a*.

In Fig. 5.2*b* a similar comparison is made for a small TE_{11}/TM_{11} dual mode horn ($a = 2{\cdot}4\ \lambda$). In this case the illumination at the aperture rim is approximately independent of observation plane.

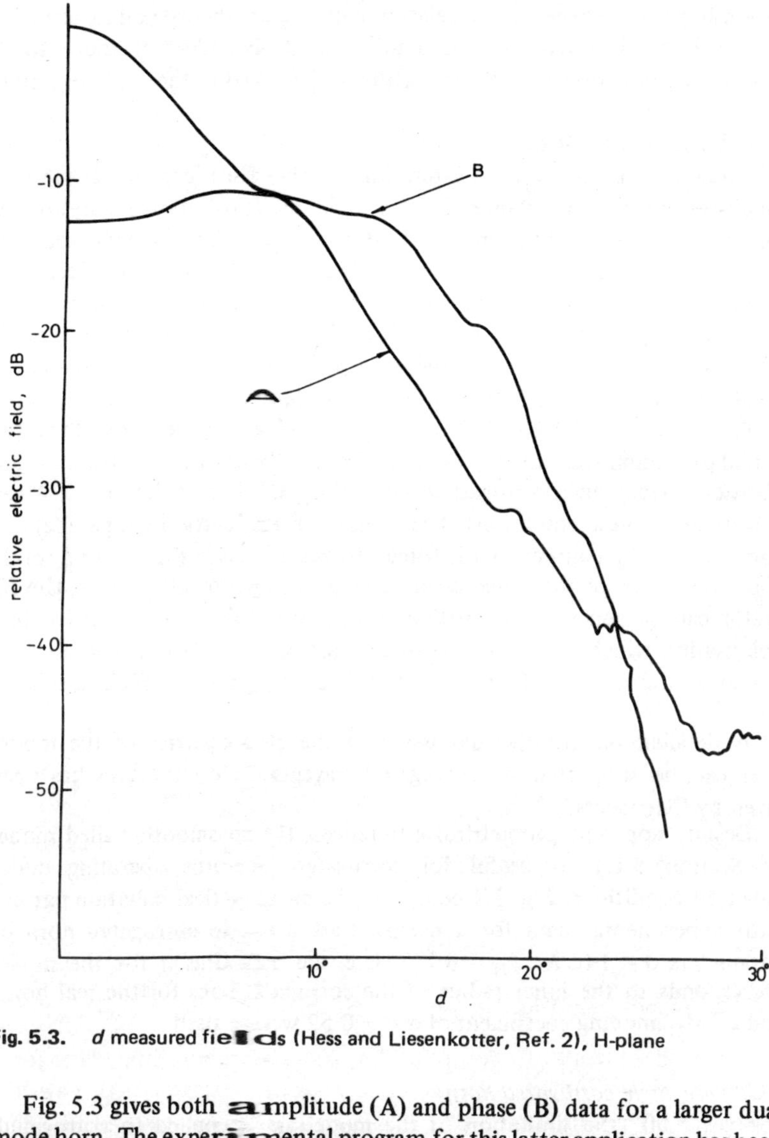

Fig. 5.3. *d* measured fields (Hess and Liesenkotter, Ref. 2), H-plane

Fig. 5.3 gives both amplitude (A) and phase (B) data for a larger dual mode horn. The experimental program for this latter application has been described in References 2 and 3, and was carried out in the MBB open-ended anechoic chamber. The fields were measured at a radius of 214λ with a centre of rotation offset 22 λ behind the aperture of the horn, and the patterns have also been computed for this finite range. The horn geometry was specified by $a = 5{\cdot}74\,\lambda, \psi_H = 9{\cdot}6°$ and the moding coefficient was $\alpha = 0{\cdot}77\angle 173°$. The larger aperture size is partly responsible for the excellent correlation between theory and experiment for this horn. The agreement is within 0·5 dB and 4° to the −40 dB point of the radiation pattern, and within 0·25 dB to the — 30 dB point.

Narrow flare angle corrugated horn

A waveguide having a corrugated inner surface can under certain conditions support a 'balanced hybrid' field distribution satisfying the conditions of eqns. 3.63 and 3.64. It then radiates a polar diagram having many advantageous properties. These include a high degree of circular symmetry and a low level of radiation at wide angles. With single mode propagation, the phase change of the guided wave is no longer a critical parameter and these properties can be maintained over a wider bandwidth (e.g. 2:1) than the smooth-walled Potter horn.

To support a balanced hybrid mode ideally requires that the boundary conditions for $\boldsymbol{E}$ and $\boldsymbol{H}$ at the surface of the structure are identical. Transverse corrugations of width $\ll\lambda$ and depth $\approx\lambda/4$ can approximate these conditions. The transverse electric field parallel to a corrugation (E_ϕ component) is forced to zero over the top of a tooth. Since no mode of this polarisation can propagate in the slot, the E_ϕ vector cannot penetrate it. Furthermore, since H_ϕ is associated with a polarisation which can propagate in the slot, a λ/4 slot throws an open circuit at the inner surface $r = a$ for this case, so that H_ϕ is also zero here.

A detailed quantitative discussion of the properties of the modes that can be supported by corrugated waveguide structures has been given by Clarricoats.[4, 5]

Despite apparent geometrical differences, the smooth-walled model of Section 5.1.1. is useful for corrugated horns operating under balanced conditions. Fig. 5.4 compares the analytical radiation pattern with experimental data for a narrow flare angle corrugated horn of parameters $a = 1{\cdot}65\,\lambda$, $\psi_H = 6{\cdot}7°$. Here the radius a for the model corresponds to the inner radius of the corrugations for the real horn, and a TM_{11} moding coefficient of $\alpha = -0{\cdot}52$ was used.

Wide flare angle corrugated horn

For $\psi_H > 50°$, the limitation of the model as applied to corrugated horns starts to become apparent (e.g. see Fig. 5.5, $\psi_H = 60°$, $a = 2{\cdot}3\,\lambda$). The discontinuity at the horn throat (junction of circular waveguide and corrugated waveguide) causes higher-order modes to be launched in addition to the wanted balanced hybrid mode.[6] For wide flare angles there is insufficient length of corrugated guide which is beyond cutoff to these higher modes to cause them to be suppressed.

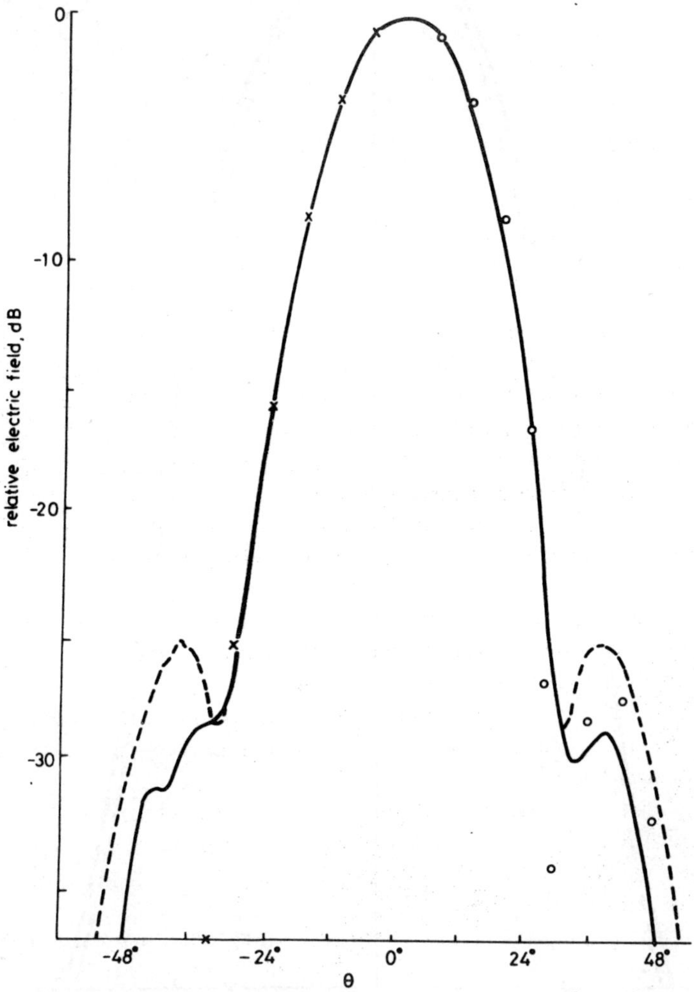

Fig. 5.4. *Comparison of measured and theoretical patterns for a transversely corrugated feed horn*

——— H-plane, measured
– – – – E-plane, measured
o H-plane, theoretical
x E-plane, theoretical

For a wide flare angle horn, it is better to measure the radiation pattern and use an S.W.E. fit for modelling as per Section 5.2.3.

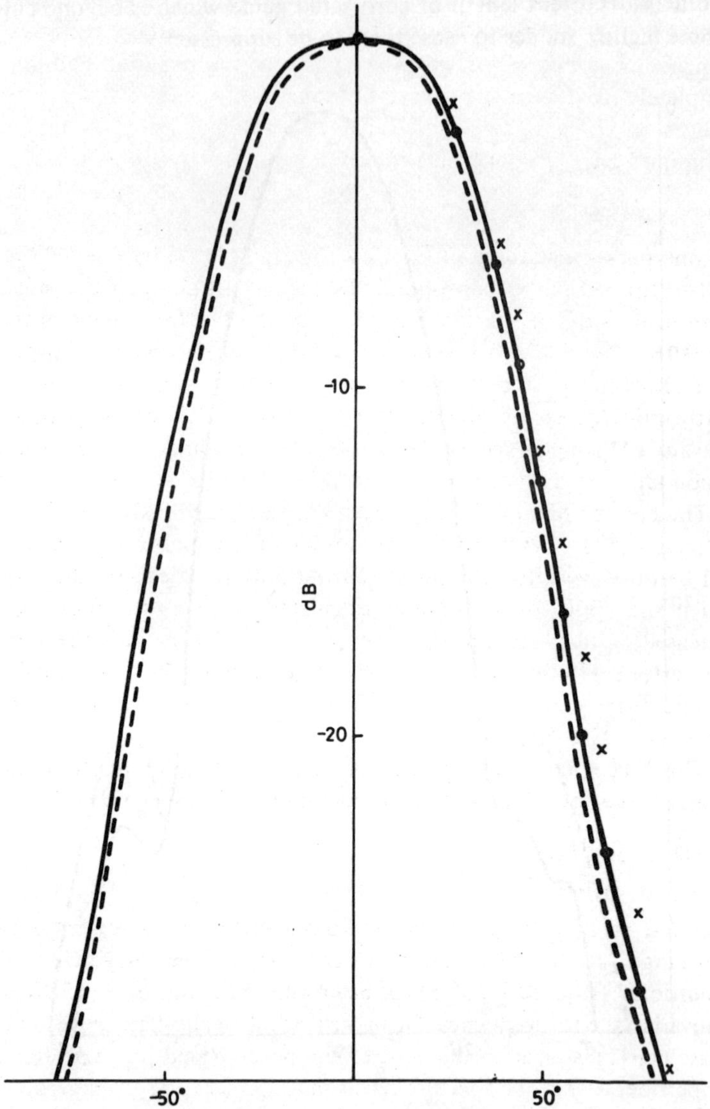

Fig. 5.5. *Radiation for a wide flare angle corrugated horn*

——— H-plane, measured

– – – – E-plane, measured

o E, theoretical

x H, theoretical

5.1.3 Phase centre for conical horns

When used as a feed for a reflector, the phase properties of a horn can become of considerable importance. In general, a conical horn does not possess a unique phase centre; that is, it is not possible to rotate the horn about a centre for which the phase of the radiation pattern ψ is completely independent of rotation angle θ. A 'best fit' phase centre reduces ψ to a minimum residual value. Sophisticated minimisation techniques can be conceived, but often it suffices to stipulate that

$$\psi(\Theta) = \psi(0) \tag{5.16}$$

for one particular value of Θ. Thus for example we have a '6 dB phase centre' for which Θ corresponds to the -6 dB point of the radiation pattern. A 6 dB centre tends to minimise the average residual phase for the 0 dB to -9 dB portion of the radiation pattern, and is appropriate to front-fed or hyperbola/parabola reflector applications. A 14 dB centre minimises the residual over the 0 dB to -17 dB pattern, and is relevant to shaped Cassegrainian systems for which this larger illumination taper is the norm.

The curves of Fig. 5.6 have been derived analytically for the model of Section 5.1.1. with $a = 3\lambda$, by varying the geometrical parameters. It will be observed that the phase centre position varies with the range at which the field distribution is calculated. Also as the flare angle is increased, the phase centre moves inwards from the aperture towards the vertex. However, it never quite reaches this point, and eventually its direction of motion is reversed and becomes in sympathy with the vertex (see Fig. 5.6*c*).

This latter occurrence marks the end of 'diffraction limited' and the onset of 'frequency-independent' operation, and is specified by[7]

$$a^2\left(\frac{1}{R_H}+\frac{1}{R}\right) = \lambda \tag{5.17}$$

Here R_H is the distance from the horn aperture to its vertex, and R is the range to the observation surface. Under frequency-independent conditions (e.g. see Fig. 5.5) the beamwidth at about the -15 dB point is invariant with frequency, being just equal to the flare angle, and the phase centre is near to the vertex. The penalty paid for these desirable properties is that the aperture size is much larger for a given beamwidth than for a narrow flare angle horn (resulting in blockage when feeding a reflector).

Some differences in the behaviour of the 6 dB and 14 dB phase centres can be discerned from Fig. 5.6. This is symptomatic of the existence of a finite residual ψ about either centre. This residual is typically $\approx 30°$ peak-to-peak for many horn geometries. In reflector antenna applications, the associated aperture illumination phase error can cause a significant degradation in performance if it happens to be additive to other phase errors such as those associated with the manufacturing tolerances for the reflector profile.

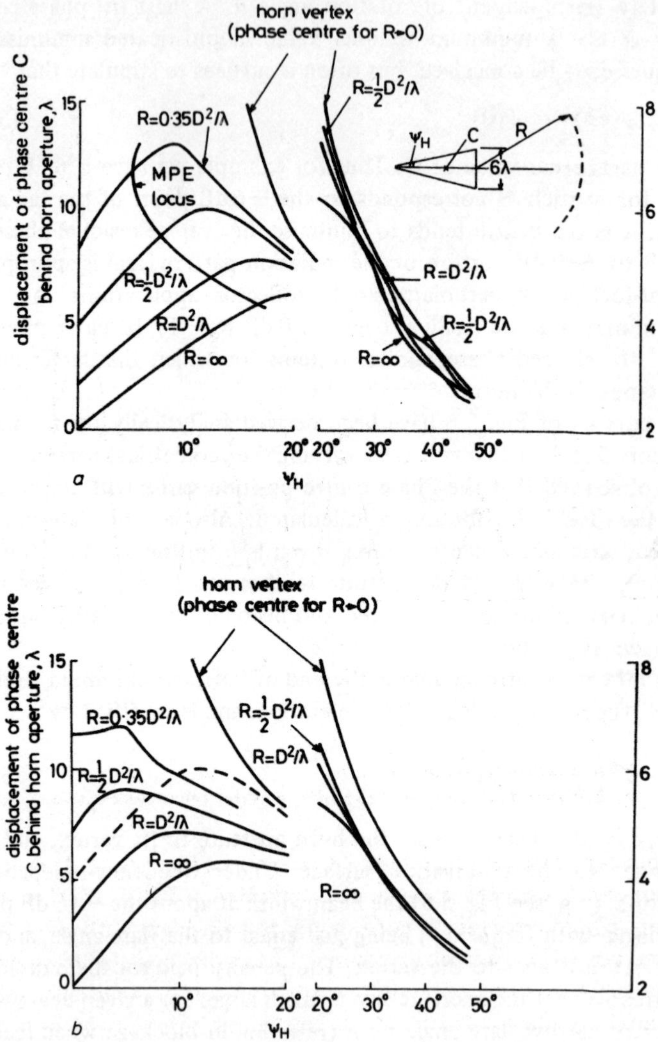

Fig. 5.6. *Phase centre for a conical horn*
a 6 dB
b 14 dB

represented in terms of the third definition[8] of Ludwig (see Chapter 6). For the analytical model of Section 5.1.1. the crosspolar level is in general a function of the TM_{11} moding coefficient. For the nominal $\alpha = -0{\cdot}52$ case, the crosspolar level is small but nevertheless always finite, being typically ≈ -35 dB.

The dual mode model therefore cannot accurately represent the crosspolar properties of a corrugated horn. An apparently more rigorous treatment for the latter would model the aperture distribution as a balanced hybrid mode. For the finite ψ_H case, this mode can be formulated in terms of Legendre and Hankel functions of noninteger[9] order n. The index 'n' in turn is determined by a transcendental equation involving the geometrical parameters.

In general, balanced hybrid modes are characterised by equal and opposite tilts of the $\boldsymbol{E}$ and $\boldsymbol{H}$ lines at any field point (see Section 3.2.7.2). It transpires that for the balanced condition the radiated pattern has precisely zero crosspolar content in the far field. In fact at the balanced hybrid mode frequency, the hybrid model is no closer to physical reality for a corrugated horn than the simple model of Section 5.1.1.

For example, Fig. 5.8 (due to R. W. Ashton) gives some measured data for a series of experimental horns, each being operated at the balanced mode frequency for its corrugations. The level of cross-polarisation is seen to be finite, although it diminishes as the size of the horn aperture is increased. Thus, the usefulness of the hybrid mode model lies not in this case but rather in that of a horn operated at a frequency well above the balanced hybrid mode frequency. The model then provides an approximate prediction of the additional cross-polarisation due to mode unbalance.

Recent work[10] has shown that the residual crosspolar pattern under balanced hybrid conditions can be accounted for analytically when the hybrid mode model is refined to include

(*a*) The conducting flange surrounding the aperture. It appears to be better to model an infinite flange to simulate the final corrugation than no flange at all. An infinite flange corresponds to omitting the electric dipole terms of the aperture radiation, see Section 2.2.5.2.
(*b*) Finite pitch corrugations. These introduce higher order Floquet harmonics in addition to the balanced hybrid fundamental space harmonic.

In practice it is often advantageous to carry out experimental measurements as described in Section 5.2., rather than attempt to introduce this degree of sophistication into the analytical model.

Although as already stated TM_{11} moding is in general necessary in a smooth-walled horn to maintain low crosspolarisation, this no longer holds good for small apertures. In particular, the dominant TE_{11} mode alone in a circular pipe of 0·7 λ diameter radiates a beam with only −48 dB crosspolar content, according to the analysis of Section 5.1.1. This is because with the pipe operated not far from cutoff, the aperture impedance is high and radiation from equivalent magnetic current content of the aperture fields exceeds that from the equivalent electric currents. In terms of the treatment of Section 6.2.3, the inherent cross-

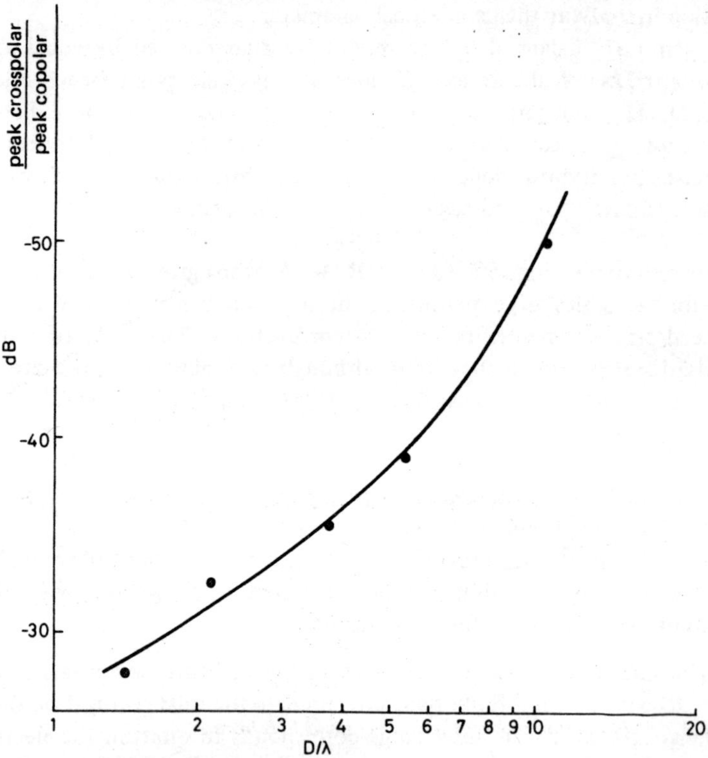

Fig. 5.8. *Variation of peak crosspolar signal for $\psi_H = 2{\cdot}5°$ corrugated horns*

polarisation of the magnetically biased elemental cell pattern cancels that of the vector space factor associated with the curved field lines of the TE_{11} mode.

Inevitably, strong external currents will be excited on an open-ended guide having a diameter as small as 0·7 λ. For an ICR or impedance controlled radiator[11] feed, these currents are suppressed by one or more peripheral chokes.[11, 12] Alternatively, the guide may radiate through an aperture in a ground plane. In this latter instance, it is appropriate to drop the *H*-field terms of eqns. 13 and 14, the radiation then being calculated from twice the *E* terms (Section 2.2.5.2). In effect, the ground plane causes the aperture to radiate *with infinite impedance.*

The elemental cell pattern thus has a very high level of cross-polarisation in the ground plane case. As the aperture size is reduced, the space factor associated with its focusing properties broadens and the product, space factor times crosspolar cell pattern, increases as for the ICR horn. However, equality with the crosspolar response due to curved field lines is now reached when the aperture size is still relatively large. Adatia[13] has demonstrated that a diameter of 1·15 λ minimises crosspolarisation with a ground plane TE_{11} radiator.

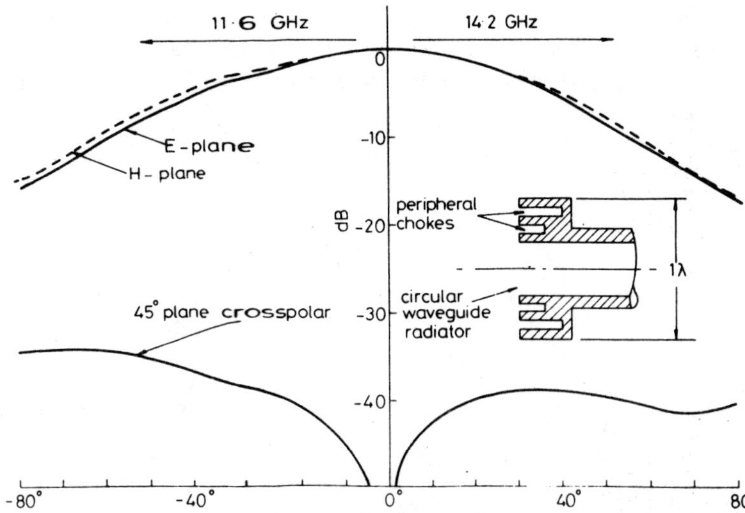

Fig. 5.9. *Radiation patterns for a dual band ICR horn*

The quarter wave chokes on an ICR feed radiate in their own right at a very high impedance. Thus for optimum performance, the inner guide diameter needs to be increased from 0·7λ to ≈ 0·8λ to introduce compensation.[11] In practice crosspolar suppression to −35 dB or −40 dB can be obtained (Fig. 5.9) while maintaining a low backlobe and minimal aperture blockage for front-feed applications.

The simple impedance principles outlined above do not extend readily to accurate quantitative modelling or design for an ICR feed. A hybrid mode treatment as for corrugated horns is not feasible, owing to the number of different modes launched. Recently, integral equation techniques have been applied.[14–16] The associated algorithms tend to be rather complex and computationally demanding, however. When analytical modelling of the far field principal lobe is all that is required, empirical results suggest the use of a simple angular scaling factor in the range 0·7 to 0·8. This factor is applied to the pattern for a 0·7 λ TE_{11} radiator calculated according to Section 5.1.1. The resulting formulation is not applicable in the null and backlobe regions of the polar diagram.

5.1.5 Modelling other types of feed system

In some types of reflector antenna a composite feed system is substituted in place of the feed horn. For example, a beam waveguide consists of a set of offset reflectors arranged[17] in the fashion of an optical periscope (Fig. 5.10). This type of feed system is becoming popular for Cassegrainian earth stations for satellite communication systems. The BWG provides a low loss signal path between the Cassegrain secondary focus and the electronic receiving equipment. It also allows the antenna to track the satellite without introducing requirements for waveguide rotary joints or an elevated equipment cabin.

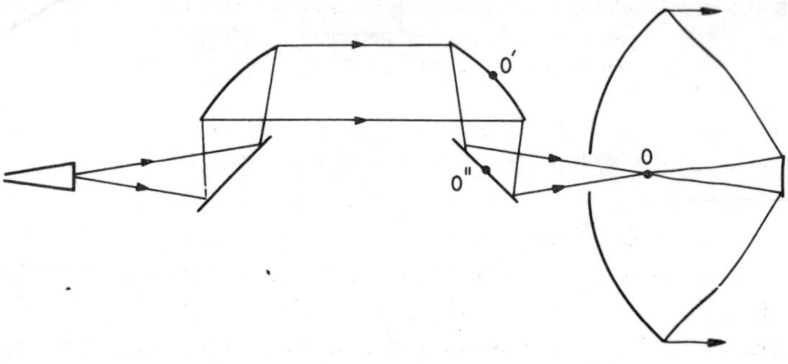

Fig. 5.10. *Schematic of beam feed Cassegrain with 2 offset parabolic and 2 plane mirrors*

A typical BWG consists of two plane and two parabolic mirrors. The plane mirrors are usually of similar projected diameter to the parabolic ones. Thus, GO ray tracing would suggest that only the central portion of each plane mirror is active. This hypothesis has been tested quantitatively using the vector diffraction methods of Chapter 3. It was found that the plane mirror scatter pattern evidences only a very minor rim diffraction effect.[18] Thus, it is permissible to replace the plane mirrors by infinite reflecting surfaces in an analytical treatment. Application of the principle of images then yields the equivalent system of Fig. 5.11*a*, having only two reflectors.

A final simplification is to replace the offset paraboloids with an 'equivalent' in line rotationally symmetric system (Fig. 5.11*b*). It is necessary to maintain the same projected aperture diameter, focal distances, and centre-to-centre spacing of reflectors. The symmetric model is well adapted to an analysis using the S.W.E. treatment of Section 3.2.7.1.

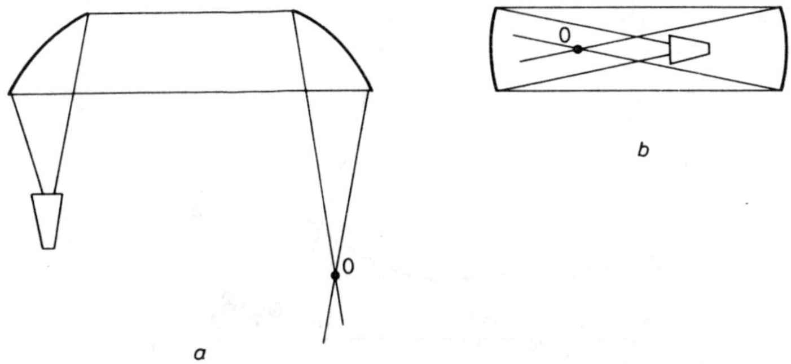

Fig. 5.11. *Simplified models for a beam waveguide feed system*
a model using 2 offset reflectors
b model using 2 symmetric reflectors

In a particular case, the primary feed horn was modelled according to Section 5.1.1. with $a = 5{\cdot}11\,\lambda$ and $\psi_H = 2{\cdot}5°$. Fig. 5.12*a* gives the computed fields at the surface of the first BWG model reflector. The latter pattern is tabulated versus off-axis angle measured at the reflector focus. The induced current distribution in the first reflector was then calculated by physical optics. This allowed the Laurent S.W.E. spectrum to be evaluated according to eqns. 3.60–3.62. Only $m = 1$ modes are excited.

The next step is to re-expand the series at the surface of the second reflector (Fig. 5.12*b*). It will be noticed that the beam has broadened somewhat at this point, and the phase deviation relative to a point source at the reflector focus has increased.

Finally, the Laurent spectrum for the induced currents at the second reflector is calculated in a similar fashion. In Fig. 5.12*c*, the corresponding radiated fields have been evaluated on a $184\,\lambda$ radius spherical surface in the vicinity of the Cassegrainian subreflector. However, it is

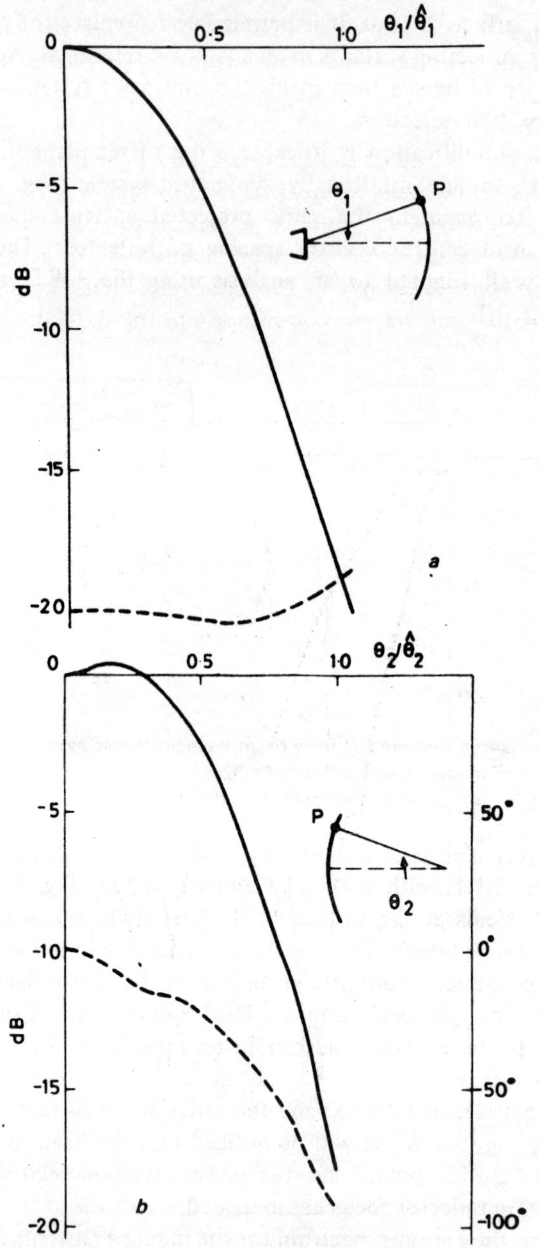

Fig. 5.12 ***Predicted field transformation in beam waveguide***
a at first reflector ——— amplitude at P
b at second reflector – – – phase at P

often convenient to represent the BWG system via the Laurent spectrum itself. The spectrum data can be 'carried over' for further processing to predict the performance of the complete Cassegrain antenna.

The approach described can provide a very powerful design tool for BWG systems.[17] For example, it can be used to optimise the geometrical parameters of the feed and reflector system. This optimisation can take into account electrical performance features such as the overall efficiency of power transfer to the Cassegrain subreflector, the field levels and noise temperature contribution at the BWG outer shielding

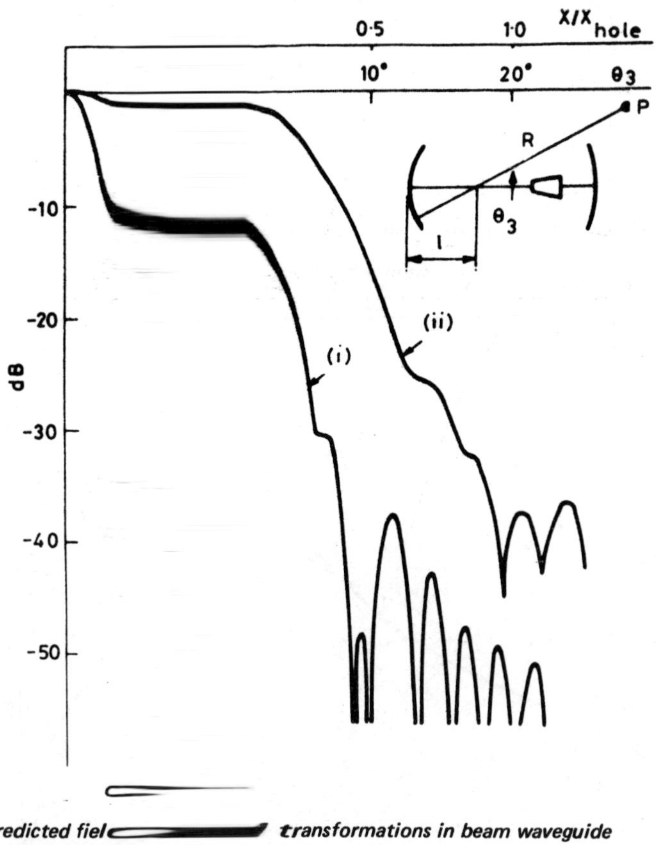

Fig. 5.12. *Predicted fiel transformations in beam waveguide*
c scattered ff second reflector

(i) at R = 184 λ, θ_3 scale
(ii) at hole in ain reflector, x/x_{hole} scale

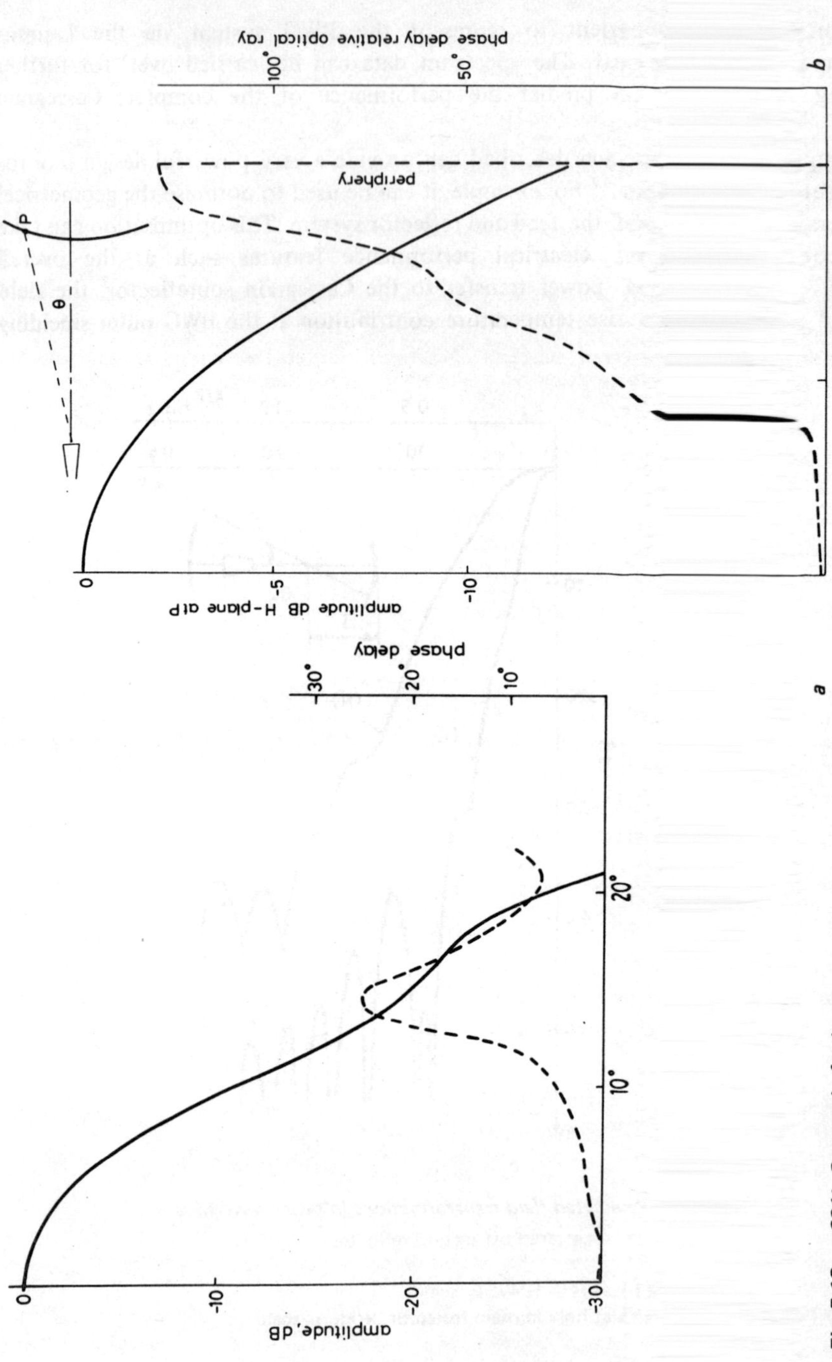

Fig. 5.13. *621λ Cassegrain feed system patterns;*
a measured H-plane feed pattern *b* field at subreflector surface
—— amplitude } 110λ sphere —— dBs
– – – phase – – – phase

tube, and efficient transmission of power through the hole in the Cassegrain main reflector.

5.2 Feed modelling with measured pattern data

5.2.1 Polynomial interpolation schemes

Many types of feed horn do not lend themselves well to analytical modelling. In particular, the small aperture pyramidal horn is difficult to model accurately.[19] The solution usually then adopted is to make use of measured pattern data. For reflector antenna calculations of either the KHS type (Chapter 2) or the S.W.E. type (Chapter 3), field values will usually be required at a closely spaced 'integration grid'. It is rarely convenient to measure feed horn polar diagrams directly on this grid.

However, the intermediate grid points may be derived by numerical interpolation, using a local parabolic fit (or second-order Aitken interpolation). Alternatively, a cubic spline[20] fit is sometimes used. For the three-dimensional case local polynomial surfaces can be specified of the form

$$z = f(x,y)$$

In one implementation,[21] terms to x^2, xy, and y^2 are fitted to the groups of nine measured points nearest to the desired interpolated point on a least squares basis.

5.2.2 Fourier interpolation

If the antenna aperture fields are expressed as a Fourier series, the radiation pattern takes the form

$$\sum_m c_m \frac{\sin \pi(u-m)}{\pi(u-m)} \tag{5.18}$$

where

$$u = D/\lambda \sin\theta \tag{5.19}$$

These equations amount essentially to a plane aperture scalar diffraction theory result. Apart from this limitation, they are valid for any aperture shape, D being the maximum aperture dimension measured parallel to the plane of cut.

Typically 3 or 4 pattern samples at unit u intervals on either side of boresight yield a sufficient number of terms for eqn. 18 to permit its use in interpolating any point on the main lobe of the cut. Thus, in relation to the number of data samples required, the method can be more efficient than polynomial interpolation. However, measured phase data extending into the sidelobe region is always required, and there is a tendency to accentuate any errors in the data.

5.2.3 Spherical wave expansion fit

An alternative to polynomial interpolation is to fit a Laurent S.W.E. to the measured data. References 22 and 23 discuss the formulation and the experimental methods appropriate in the context of a general technique applicable to arbitrary radiation patterns and an automated near field test range. For a feed horn, the technique can often be simplified, both in terms of the number of data samples to be measured and the number of modes to be included in the expansion. Nevertheless, to ensure a sufficiently accurate evaluation of the S.W.E. weights, pattern measurements must be taken over a somewhat larger solid angle than that associated with the capture area of the reflector that the horn is intended to illuminate. Thus amplitude and phase data extending to the −30 dB or −35 dB point of the pattern will ordinarily be needed. However, substantial advantages accrue from an S.W.E. fit in that

(*a*) The data is fitted to an analytical expansion which is physically compatible with the aperture dimensions involved. Random errors for individual data points are effectively smoothed.
(*b*) The S.W.E. removes the inconvenient restriction that the feed fields should be measured at a radius closely corresponding to the feed/reflector separation.
(*c*) The Laurent spectrum comprises a unique representation of the measured field distribution. It becomes a standard data interface available to link with any reflector analysis or synthesis software accommodating this type of input.

As an example, Fig. 5.13*a* gives measured amplitude and phase patterns for a feed horn for a 621 λ Cassegrain antenna. The horn is of the corrugated conical type and hence only E- and H-plane patterns are necessary to characterise the radiation properties (see Section 2.2.11). The measurements were carried out at a range of 110 λ from the horn aperture. A Laurent spectrum was fitted to this data according to eqns. 3.90–3.95. The maximum radial index n_L for the expansion was 25, $m = 1$ modes being excited.

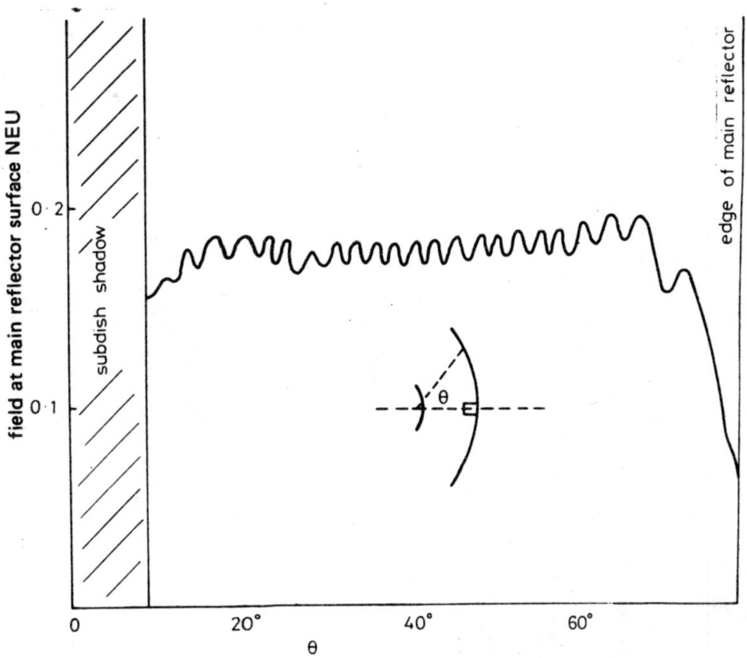

Fig. 5.14 *Field at main reflector for 621 λ Cassegrain*

The subreflector for the Cassegrainian system lies at a range of 150 λ (vertex) to 168 λ (rim). The fields at the subreflector surface are obtained by re-expanding the Laurent S.W.E. and are illustrated in Fig. 5.13*b*. The phase of the subreflector illumination is plotted 'relative to ray optics', e.g. relative to that of an ideal point source placed at the centre of the feed aperture. It will be observed that the phase pattern has changed significantly in relation to the phase at the measurement range.

The next stage of the calculation is the evaluation of the subreflector scatter pattern at finite range on the surface of the main reflector, via a Laurent S.W.E. as per eqns. 3.60–3.62. This scatter pattern is presented in Fig. 5.14; the units are NEU, defined in the footnote at the beginning of Chapter 2. The induced currents on the main reflector are then calculated according to the principle of physical optics. The main reflector secondary radiation pattern of Fig. 5·15 takes into account a central circular shadow due to the subreflector, and is calculated using the KHS diffraction formulation of eqns. 2.35–2.41.

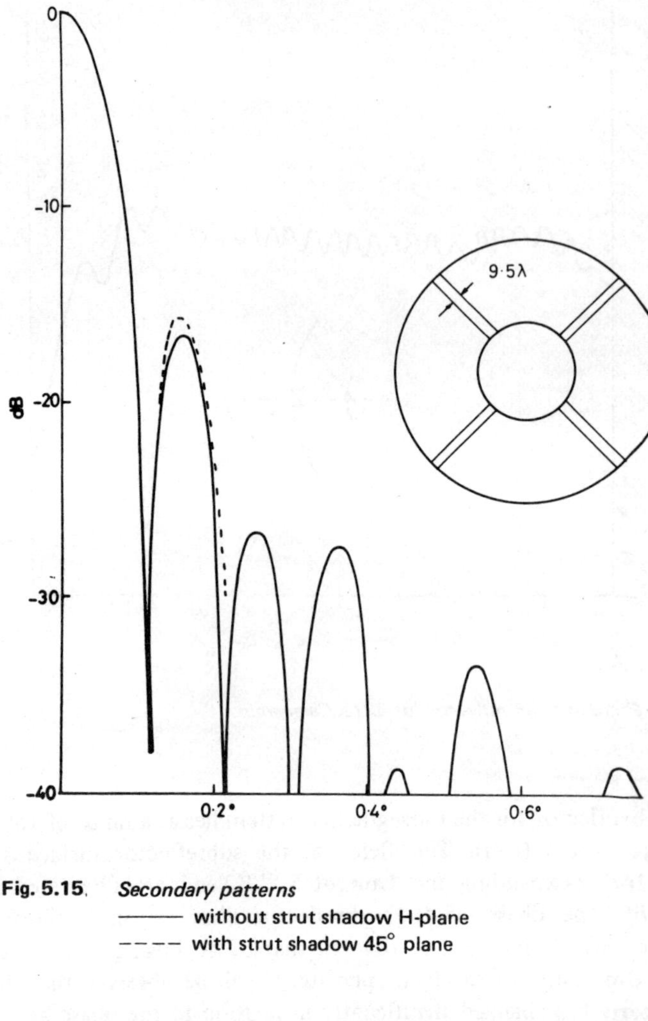

Fig. 5.15 *Secondary patterns*
——— without strut shadow H-plane
– – – – with strut shadow 45° plane

For completeness, the secondary pattern after incorporating the additional blockage shadow due to the subreflector support struts is also included as the broken curve in Fig. 5.15

References

1 POTTER, P. D.: 'A new horn antenna with suppressed sidelobes and equal beamwidths', *Microwave J.*, 1963, pp. 71–78

2 HESS, W., and LIESENKOTTER, B.: 'A dual mode feed horn for deep space

Chapter 6

Crosspolarisation in reflector antennas

6.1 Fundamental principles

6.1.1 Reference vectors

The crosspolar response of an antenna may be defined as its response to a field polarised orthogonally to a reference, 'copolar' field. This type of definition of course begs the question in that the reference field is left unspecified. In fact in the general case it will be necessary to specify

(*a*) a reference *z*-axis (e.g. nominal boresight) and a reference origin (e. g. antenna phase centre)
(*b*) a reference polarisation at this axis orthogonal to it (e.g. copolar vector on boresight); in general the unit vector associated with the copolar reference polarisation may be represented in the form

$$(\boldsymbol{a}_{co})_{\theta=0} = \boldsymbol{a}_x \cos\phi_0 + \boldsymbol{a}_y \sin\phi_0 \tag{6.1}$$

where ϕ_0 is the inclination of the reference polarisation to the x-axis
(*c*) a definition relating $\boldsymbol{a}_{co}$ at off-axis incidence to the on-axis case of eqn. 1 above.

Frequently (but not always) the antenna will possess geometrical symmetries which lead naturally to a unique choice for the reference conditions (*a*) and (*b*). The question of a definition for the off-axis case is less readily diposed of, however. In this book the definition adopted will be the '3rd definition' of Ludwig.[1] Thus

$$\boldsymbol{a}_{co_3} = \boldsymbol{a}_\theta \cos(\phi-\phi_0) - \boldsymbol{a}_\phi \sin(\phi-\phi_0) \tag{6.2}$$

and

$$\boldsymbol{a}_{x_3} = \boldsymbol{a}_\theta \sin(\phi-\phi_0) + \boldsymbol{a}_\phi \cos(\phi-\phi_0) \tag{6.3}$$

The principal advantage of this scheme is that it tends to preserve the polarisation purity of a field distribution through a series of reflectors. Thus a focused reflector antenna forms a narrow beam in the far field, and the crosspolar characteristics of this beam are fairly insensitive to which of Ludwig's off-axis definitions are used. However, the primary feed pattern and subdish scatter pattern extend to a fairly wide angle. A primary pattern which has the vector properties of $\boldsymbol{a}_{co3}$ will result in a copolarised far field pattern to the order of GO provided that the reflector geometry satisfies certain elementary conditions to be discussed in Section 6.2.1., By way of contrast, if the primary feed pattern is specified according to one of Ludwig's alternative definitions, the far field pattern will be substantially depolarised by any definition.

Another advantage of eqns. 2 and 3 is that if two orthogonally polarised sources with collinear boresights each launch a copolarised field, then the two radiated field vectors will be orthogonal at all points in space. This property is closely related to the requirements for 'frequency re-use' communications sytems.[2]

For circular polarisation, copolar (say RHCP) and crosspolar (say LHCP) may be distinguished unambiguously in the off-axis case without need of any special definition. However to determine the phase angle associated with a CP pattern, a reference vector is required. Thus specifications (*a*), (*b*) and (*c*) as discussed earlier remain applicable to the CP case.

When the reference vector is specified according to definition 3, the 'wanted hand' and 'unwanted hand' CP unit vectors become

$$\boldsymbol{a}_{wcp} = \frac{1}{\sqrt{2}}(\boldsymbol{a}_{co_3} + j\boldsymbol{a}_{x_3}) \tag{6.4}$$

and

$$\boldsymbol{a}_{ucp} = \frac{1}{\sqrt{2}}(\boldsymbol{a}_{co_3} - j\boldsymbol{a}_{x_3}) \tag{6.5}$$

For practical antennas, the phase of the radiation pattern according to this definition tends to remain sensibly constant over a focused CP beam.

For the remainder of this book, a '3' suffix will not be quoted explicitly but will be understood.

6.1.2 A special case–pseudo-circular symmetry

An important class of antennas launch field distributions having pseudo-circular symmetry (see Section 2.2.11). Consider a field distribution

$$f_1(\theta)\cos\phi\boldsymbol{a}_\theta - f_2(\theta)\sin\phi\boldsymbol{a}_\phi \tag{6.6}$$

This vector pattern can also be written in the form

$$\{f_1(\theta)\cos^2\phi + f_2(\theta)\sin^2\phi\}\{\boldsymbol{a}_\theta\cos\phi - \boldsymbol{a}_\phi\sin\phi\}$$
$$+\tfrac{1}{2}\sin 2\phi\{f_1(\theta) - f_2(\theta)\}\{\boldsymbol{a}_\theta\sin\phi + \boldsymbol{a}_\phi\cos\phi\} \qquad (6.7)$$

Thus the crosspolar pattern constitutes one half of the difference of the E- and H-plane patterns, and maximises in 45° planes where $\sin 2\phi = \pm 1$. Zero crosspolarisation implies complete rotational symmetry of the primary pattern.

6.1.3 Electric and magnetic elementary sources

The polarisation pattern for the far field of an element of electric current is discussed in Section 2.2.8. For an element parallel to the x-axis, ignoring constants of proportionality,

$$\boldsymbol{E}_{el} = (\sin^2\phi + \cos^2\phi\cos\theta)\boldsymbol{a}_{co} - \tfrac{1}{2}\sin 2\phi(1-\cos\theta)\boldsymbol{a}_{\times} \qquad (6.8)$$

For an element of magnetic current parallel to the y-axis the corresponding result is

$$\boldsymbol{E}_{mag} = (\cos^2\phi + \sin^2\phi\cos\theta)\boldsymbol{a}_{co} + \tfrac{1}{2}\sin 2\phi(1-\cos\theta)\boldsymbol{a}_{\times} \qquad (6.9)$$

Thus either type of elementary source radiates substantial 45° plane crosspolarisation at large angles off-axis, the cross and copolar patterns being of equal level at $\theta = 90°$. An equally weighted combination of the two sources corresponds physically to a Huygens source radiator, and the ideal polarisation pattern of eqn. 2.

A circularly polarised electric current element results from combining an x-directed electric element with a similar element but y directed and in phase quadrature. It radiates a field proportional to

$$\boldsymbol{E}_{cpel} = (1+\cos\theta)\boldsymbol{a}_{wcp} - \exp(2j\phi)(1-\cos\theta)\boldsymbol{a}_{ucp} \qquad (6.10)$$

The similarity to eqn. 8 will be evident. The $1+\cos\theta$, $1-\cos\theta$ form for the 45° plane patterns of eqn. 8 is peculiar to definition 3, whereas for the CP case it obtains regardless of definition. Thus, the linear and CP cases are compatible only when definition 3 is adopted.

6.2 Front-fed reflectors

6.2.1 Polarisation properties of a GO scattering process

The vector diffraction theory principles of Chapters 2 and 3 when implemented via suitable numerical techniques are capable of yielding the crosspolar as well as copolar patterns of an antenna. However, many useful concepts concerning polarisation in reflector antennas can be derived more directly. Fig. 6.1 depicts a reflector illuminated by a primary source at 0′. According to the LPWR (local plane wave reflection) theorem of Section 2.2.6, the scattered secondary wavefront just above the reflector is specified by

$$\boldsymbol{E}_s = 2(\boldsymbol{E}_{inc} \cdot \boldsymbol{n})\boldsymbol{n} - \boldsymbol{E}_{inc} \tag{6.11}$$

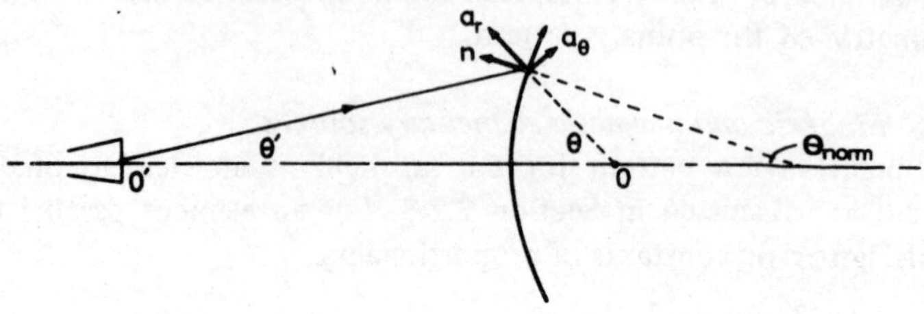

Fig. 6.1. *Illustrating reflector scattering according to the principles of geometrical optics*

Assuming a copolarised incident field

$$\boldsymbol{E}_{inc} = f(\theta')\{\cos\phi' \boldsymbol{a}_{\theta'} - \sin\phi' \boldsymbol{a}_{\phi'}\} \tag{6.12}$$

and a surface of revolution reflector having a unit normal $\boldsymbol{n}$ of

$$\boldsymbol{n} = \sin\psi \boldsymbol{a}_{\theta'} - \cos\psi \boldsymbol{a}_{r'} \tag{6.13}$$

where

$$\psi = \theta' + \theta_{norm} \tag{6.14}$$

then eqn. 11 reduces to

$$\boldsymbol{E}_s/f(\theta') = -\cos 2\psi \cos\phi' \boldsymbol{a}_{\theta'} + \sin\phi' \boldsymbol{a}_{\phi'} - \sin 2\psi \cos\phi' \boldsymbol{a}_{r'} \tag{6.15}$$

The O' co-ordinate system unit vectors may now be transformed to the 'O' system (Fig. 6.1) according to

$$\boldsymbol{a}_{\theta'} = \cos\gamma \boldsymbol{a}_\theta + \sin\gamma \boldsymbol{a}_r$$

$$\boldsymbol{a}_{r'} = \sin\gamma \boldsymbol{a}_\theta - \cos\gamma \boldsymbol{a}_r$$

where

$$\gamma = \theta + \theta'$$

Thus

$$\boldsymbol{E}_s/f(\theta') = -\cos(2\psi - \gamma)\cos\phi \boldsymbol{a}_\theta + \sin\phi \boldsymbol{a}_\phi + \sin(2\psi - \gamma)\cos\phi \boldsymbol{a}_r \tag{6.16}$$

According to this result, the condition that the θ and ϕ vector components of the scattered secondary wavefront should be copolarised according to Ludwig's third definition becomes simply not aligned with respect to a reference vector in an ideal fashion, and/or because of the crosspolarisation associated with an electric current element pattern.

6.2.4 *Diffraction depolarisation*

Front-fed parabolic reflector

Reflector depolarisation may be defined as the crosspolar response introduced by a reflector which is illuminated by a copolarised primary feed pattern.

Section 6.2.2 has demonstrated that the transverse current then flows in parallel straight lines on a paraboloid reflector. This is of course strictly true only for the focused case. Nevertheless, an axial feed displacement does not usually introduce significant induced current polarisation distortion until the reflector is defocused to the extent that there is no longer a dominant axial lobe in the far field. Thus, the transverse current space factor for a circularly symmetric paraboloid may be regarded as a scalar quantity. Vectors enter via the electric element crosspolar pattern (eqn. 8), and via axial current crosspolar radiation proportional to

$$\sin\theta \sin\phi$$

The 'focused beam' properties of the space factor ensure that the overall crosspolar pattern will maximise at a relatively small angle off-axis $\hat{\theta}_x$. For reflector diameters of a few wavelengths or more

$$\sin\hat{\theta}_x > (1 - \cos\hat{\theta}_x)$$

Axial currents then represent the dominant depolarisation mechanism. They are distributed on the reflector surface as $\cos\phi$ (see eqn. 23), and the axial current space factor will also have this 'monopulse-type' ϕ dependence. Overall, axial currents yield a $\sin 2\phi$ depolarisation pattern with four major lobes in 45° planes in phase quadrature with the copolar pattern.

Symmetrical reflector depolarisation can be treated more quantitatively in respect of both axial and transverse mechanisms, by invoking the vector diffraction methods of Chapter 2 or 3, assuming an ideal copolarised primary feed. For the curves of Fig. 6.2, the feed copolar characteristic has been modelled on that for a typical measured pattern, with −16 dB illumination taper at the reflector rim.[3] The depolarisation lobe has an absolute level of 8 dB below isotropic, independent of reflector diameter D. It is only weakly dependent on edge illumination. Relative to the copolar pattern, depolarisation vanishes in the optical limit $\lambda/D \to 0$.

Even for a paraboloid as small as 10 λ diameter, depolarisation will be at −37 dB relative to the on-axis copolar response, a level sufficient to be of significance only in a critical frequency re-use system.[2] However, special considerations apply when the paraboloid is deformed slightly to generate a shaped beam.[2,4] Thus the −8 dB absolute level tends to survive most forms of beam-shaping, even when these reduce the copolar gain of the antenna by ≈ 10 dB. For a typical 'earth coverage from a satellite' application, the within coverage worst-case crosspolar/copolar ratio due to symmetrical reflector depolarisation is of the order of −30 dB. This could represent a fundamental limitation to the implementation of frequency re-use with shaped beam reflectors.

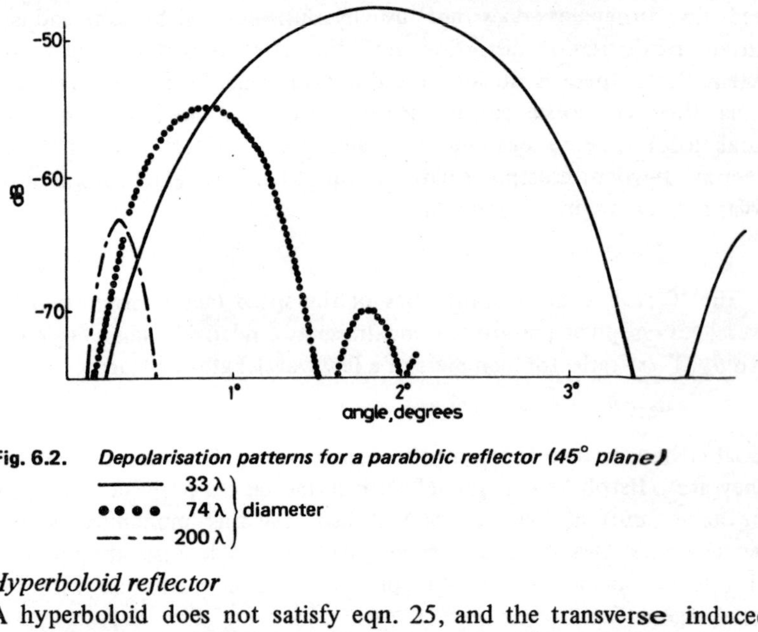

Fig. 6.2. ***Depolarisation patterns for a parabolic reflector (45° plane)***
——— 33 λ
•••• 74 λ diameter
— - — 200 λ

Hyperboloid reflector

A hyperboloid does not satisfy eqn. 25, and the transverse induced currents are curved. Moreover, the vector space factor is now a wide angle pattern which does little to suppress the crosspolar radiation of individual current elements. As a result, reflector depolarisation is a more significant effect than for a paraboloid.

Nevertheless, some degree of cancellation between space factor and element crosspolarisation occurs for large hyperboloids. The cross-polar lobe is again proportional to λ/D. The reasoning that suggests that depolarisation vanishes in the optical limit $\lambda/D \rightarrow 0$ has already been presented in Section 6.2.1. A typical depolarisation contour plot for a 5 λ hyperboloid is presented in Fig. 6.3.

the present discussion is the behaviour of the polarisation ellipse near the peak of the spill-over lobe. Thus both theory and experiment demonstrate the presence of an axial ratio of approximately 9 dB at 46° off-axis. This is almost entirely due to diffraction depolarisation in either case. Thus the analytical prediction assumes the feed launches pure circular polarisation, whilst in the experiment the polariser had an axial ratio performance better than 0·2 dB.

With the ever increasing number of geostationary orbit communication systems, intersystem interference has become more likely. In consequence, stringent requirements have arisen for the suppression of earth station wide angle sidelobes. In the case of a Cassegrainian feed, a major contribution relative to the type of envelope suggested by the CCIR arises from subdish spill-over. Thus the depolarisation at 46° off-axis in Fig. 6.4 increases the absolute signal level by over 3 dB in the region of the radiation pattern which is most critical in relation to wide angle sidelobe specifications.

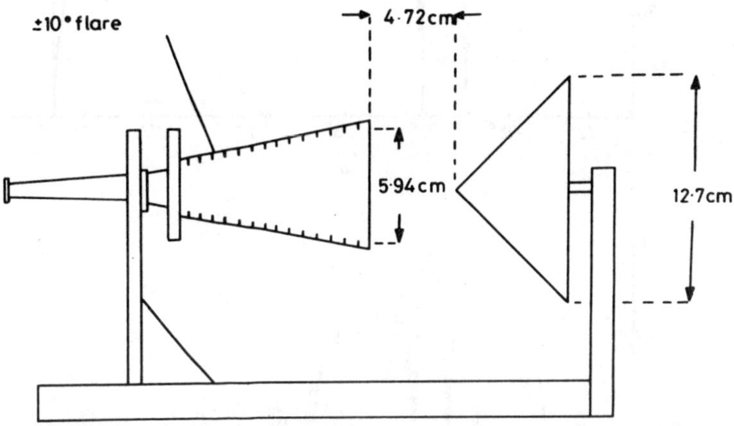

Fig. 6.5. *Horn and conical reflector*

A depolarisation experiment using a conical reflector

The depolarisation effects described above were predicted using the physical optics principle. The LPWR or 'aperture field' implementation of Section 2.2.6. would predict zero depolarisation for a perfectly focused paraboloid. Equally, the basic GTD could not model an effect which does not vary in a linear fashion with edge illumination.

PO, LPWR, and GTD are all forms of high-frequency approximation to a vector diffraction problem. Experiments to investigate the alternative hypotheses tend to be difficult to carry out since any intrinsic depolarisation effects are frequently masked by feed horn excited crosspolarisation, blockage, and test range multipath. The corrugated conical horn/conical reflector assembly of Fig. 6.5 has been chosen to provide substantial axial current radiation with an easily realisable geometry. The small diameter (5 λ) and sharp tip of the reflector also give a very stringent test of the PO assumption.

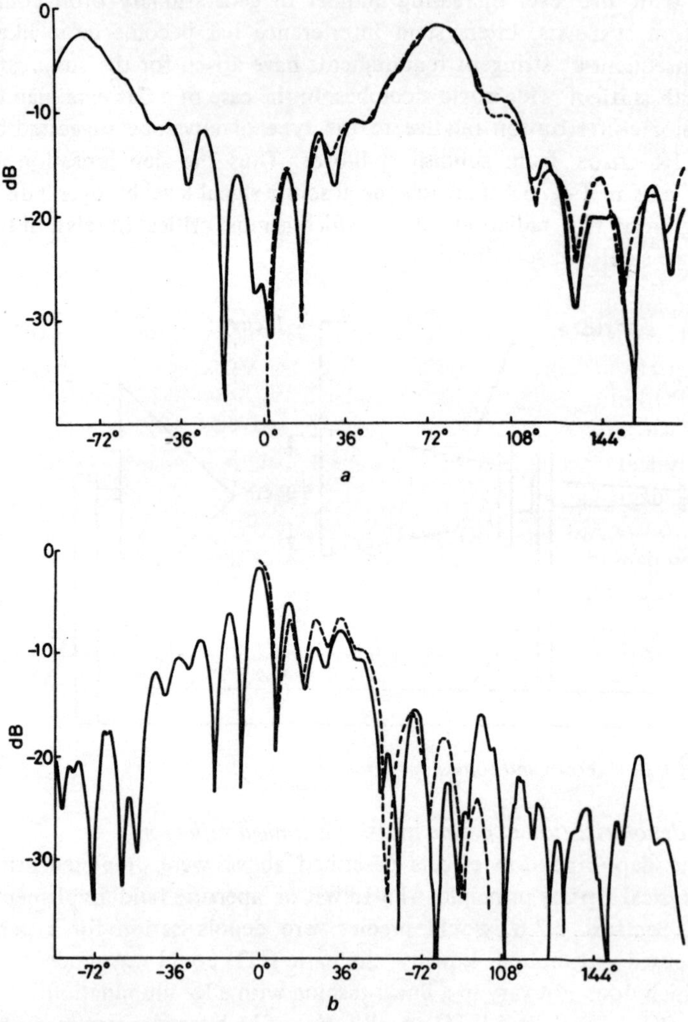

Fig. 6.6. *Radiation patterns for a conical reflector (45° plane)*
a copolar ——— measured
b crosspolar - - - - predicted

The analysis uses the feed horn model of Section 5.1.1., and Laurent S.W.E.s to represent both the feed horn and subdish radiation. As for the Cassegrainian feed system discussed above, both expansions are centred on the horn aperture to permit a direct superposition. Patterns are calculated at 6 m. range corresponding to the anechoic chamber measurement set-up. The measurement frequency is 11·8 GHz.

The chart recorder boresight for Fig. 6.6 corresponds to the centre of the rear hemisphere where the feed horn is shadowed by the reflector. 'Copolar' and 'crosspolar' are specified according to Ludwig's third definition relative to a 180° chart setting nominal case. On the basis of GO, reflected copolar energy would be bounded by 59·4° and 90° chart ordinates.

The theoretical calculations allow the overall levels of the horn + reflector patterns to be specified relative to the horn copolar pattern. The latter has been measured by removing the reflector. Thus the curves of Fig. 6.6 are indicative of the absolute as well as the relative comparison obtained between theory and measurement.

The overall agreement demonstrated, although to some extent adversely affected by chamber reflections, suffices to verify the validity of PO/induced current calculations in respect of symmetric reflector depolarisation. Feed horn blockage was not accounted for in the analytical model, and may cause the alternative lobe enhancements and depressions towards the 180° end of the chart recording in Fig. 6.6*a*. Equally, reflector support scattering may explain some of the discrepancy between the solid and broken curves in the 0° to 50° region of Fig. 6.6*b*.

6.2.5 Crosspolar conversion efficiency

The treatment so far has assumed that the feed pattern is ideal in terms of its polarisation properties. A real feed inevitably launches a finite crosspolar pattern. If $F_{\times}^{pri}(\theta)$ denotes this pattern at constant radius in the 45° plane, then the efficiency of conversion to the 45° plane secondary crosspolar pattern $F_{\times}^{sec}(\Theta)$ is defined as[5]

$$\eta_c = \left|\frac{F_{\times}^{sec}(\Theta)}{F_{co}^{sec}(\Theta)}\right|_{\Theta=\hat{\Theta}} \times \left|\frac{F_{co}^{pri}(\theta)}{F_{\times}^{pri}(\theta)}\right|_{\theta=\hat{\theta}} \tag{6.26}$$

Here $\hat{\theta}$ and $\hat{\Theta}$ are the off-axis angles corresponding, respectively, to the maxima of the primary and secondary crosspolar patterns.

In the special case of the focused paraboloid , the latter equation may readily be evaluated as per Section 2.2.11. In that analysis, the feed pattern was specified at the surface of the reflector by $f_1(\theta)$ and

$f_2(\theta)$. Now assume that the reflector lies in the far field region for the feed horn. Then extrapolation between the reflector surface and constant radius involves an $\exp(-jkr)/r$ variation, and from eqn. 7,

$$f_1(\theta) + f_2(\theta) = \frac{2r_0}{r} \exp(-jkr) F_{co}^{pri} \qquad (6.27)$$

and

$$f_1(\theta) - f_2(\theta) = \frac{2r_0}{r} \exp(-jkr) F_{X}^{pri} \qquad (6.28)$$

Reflector depolarisation effects can be mathematically suppressed by taking the limit $\Theta \to \pi$ in eqns. 2.35–2.40. Barring a constant of proportionality of $\pi r_0 \exp(-j2kF)$, these equations then reduce in the far field case to

$$I_1 = \int (J_0 F_{co}^{pri} + J_2 F_X^{pri}) r \sin\theta d\theta = I_A + I_B$$

$$I_2 = 0$$

$$I_3 = \int (J_0 F_{co}^{pri} - J_2 F_X^{pri}) r \sin\theta d\theta = I_A - I_B$$

According to eqn. 2.41, this specifies the vector properties of the far field relative to the $R\Theta\Phi$ co-ordinate system. However, the $\Theta = 0$ axis for this system is of reverse sense in relation to a forward propagating far field principal lobe. The transformation to forward sense unit vectors is

$$\Phi_f = -\Phi$$

$$\Theta_f = \pi - \Theta$$

$$\boldsymbol{a}_{\Theta_f} = -\boldsymbol{a}_\Theta$$

$$\boldsymbol{a}_{\Phi_f} = -\boldsymbol{a}_\Phi$$

Thus the far field is given by

$$E = -\{(I_A + I_B)\cos^2\Phi_f + (I_A - I_B)\sin^2\Phi_f\}\boldsymbol{a}_{co} - \sin 2\Phi_f I_B \boldsymbol{a}_X \qquad (6.29)$$

where

$$\boldsymbol{a}_{co} = \cos\Phi_f \boldsymbol{a}_{\Theta_f} - \sin\Phi_f \boldsymbol{a}_{\Phi_f} \qquad (6.30)$$

$$\boldsymbol{a}_X = \sin\Phi_f \boldsymbol{a}_{\Theta_f} + \cos\Phi_f \boldsymbol{a}_{\Phi_f} \qquad (6.31)$$

From eqn. 29 it may be deduced that the secondary crosspolar pattern again maximises in the 45° plane, and

$$\frac{F_X^{sec}(\Theta)}{F_{co}^{sec}(\Theta)} = \frac{I_B}{I_A} = \frac{\int F_X^{pri}(\theta)J_2(kr\sin\theta\sin\Theta)r\sin\theta d\theta}{\int F_{co}^{pri}(\theta)J_0(kr\sin\theta\sin\Theta)r\sin\theta d\theta} \quad (6.32)$$

Fig. 6.7 illustrates the transformation between primary and secondary crosspolar patterns for a 50λ, 72° opening semiangle paraboloid. A value of −5 dB for η_c may be deduced, although for other shapes of

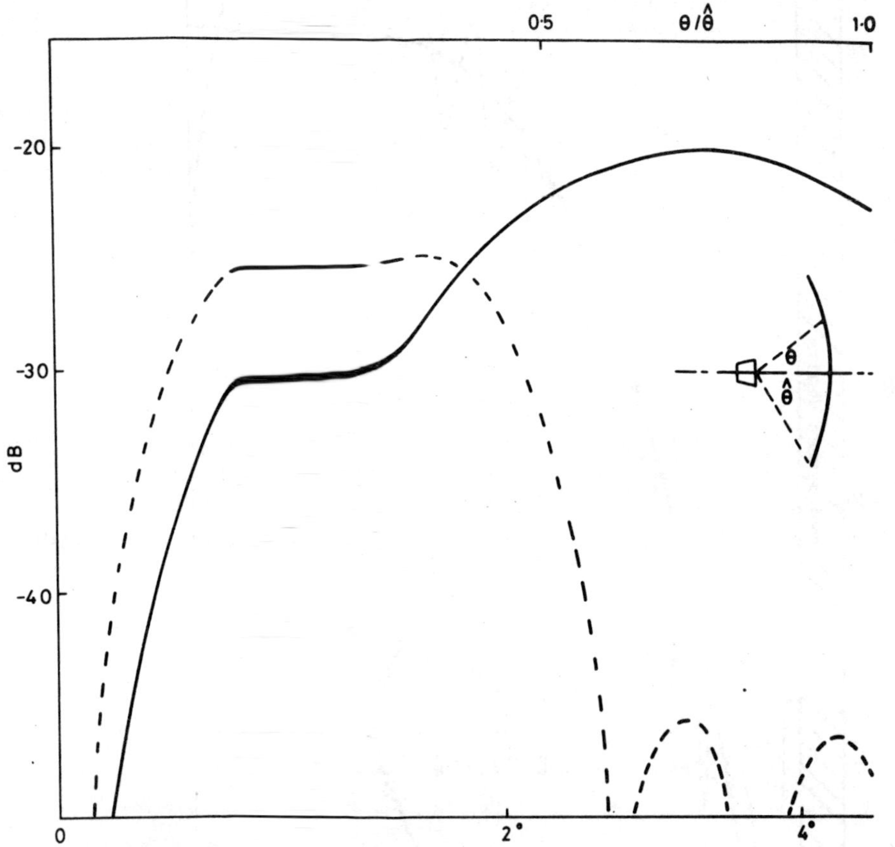

Fig. 6.7. *Crosspolar field transformations for a 50λ 72° opening semiangle paraboloid*

——— primary pattern crosspolar response on far field sphere relative on axis response, $\theta/\hat{\theta}$ scale

- - - - secondary crosspolar response relative on axis copolar

crosspolar pattern η_c may often be as low as −7 dB. η_c is essentially a real quantity. That is, the relative phase of the (primary pattern generated) secondary crosspolar pattern is the same as that obtaining in the primary pattern.

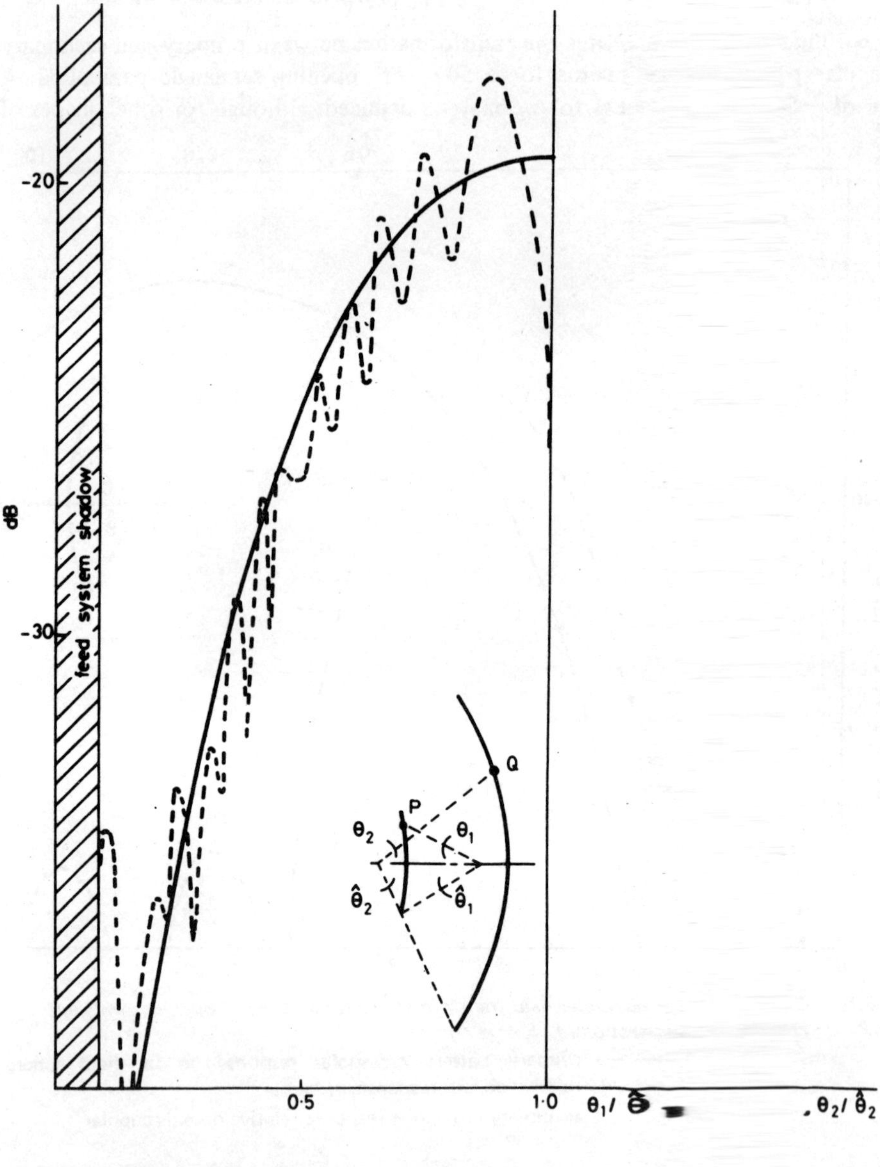

Fig. 6.8. *Crosspolar field transformations for a 40 λ hyperboloid (1 ... 3° and 75° opening semiangles)*

——— crosspolar at P relative copolar

- - - - crosspolar at Q relative copolar

Eqn. 32 may readily be evaluated by numerical integration using a computer terminal or programmable calculator. However for non-paraboloidal shapes, or near field calculations, the general vector diffraction methods of Chapter 2 or 3 are more appropriate. Two cases

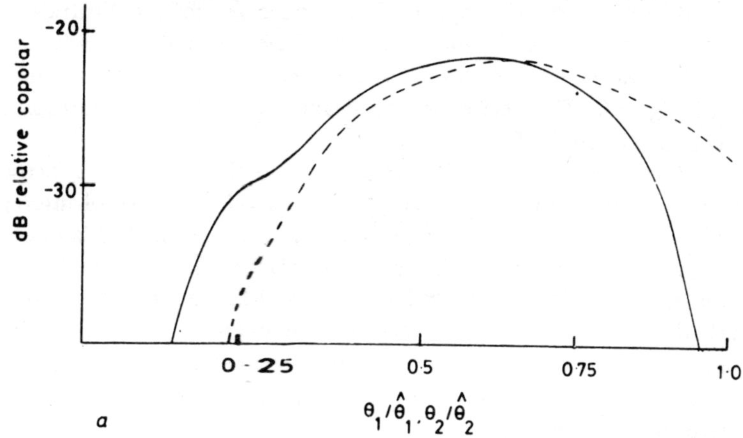

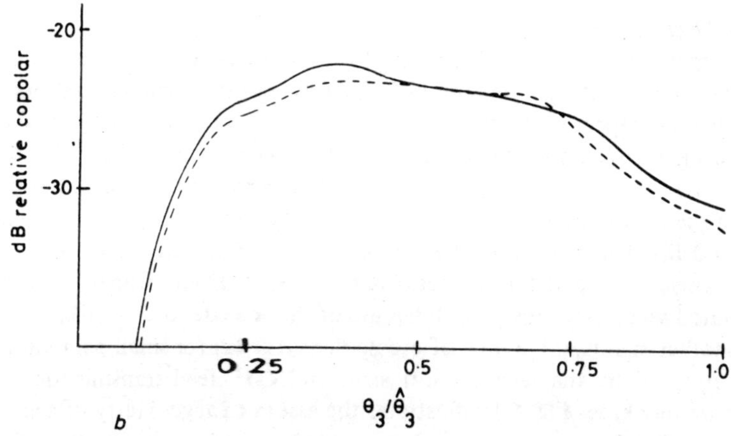

Fig. 6.9. *Crosspolar field transformations in a beam waveguide system*

a 45° plane crosspolar pattern at first and second reflectors of beam waveguide model

——— 1st reflector

- - - - 2nd reflector

b 45° plane pattern at 207 λ range from second reflector of beam waveguide model

——— symmetric model, Fig. 5.11*b*

- - - offset model, Fig. 5.11*a*

then need to be considered: reflector with real (imperfect) feed, and reflector with feed having a suppressed crosspolar pattern. By subtracting scattered patterns, the reflector depolarisation part of the solution can be eliminated.

Typical results for a 40 λ hyperboloid subreflector are illustrated in Fig. 6.8. Another example where the feed generated crosspolar lobe needs to be traced through several successive scattering operations is given in Fig. 6.9. This relates to the beam waveguide application as described in Section 5.1.5.

According to eqn. 32, η_c will always be less than unity for a focused paraboloid on the basis that $J_2 < J_0$ in the small Θ range of interest. However, there can be special circumstances under which this rule no longer applies. The elliptical beam application discussed in Section 6.3.2. is one example. Here $\eta_c \approx 1{\cdot}1$, principally because of the profile perturbation introduced to broaden the beam in one plane.[6] The resultant defocusing effect for the copolar beam finds no parallel for crosspolarisation since the 45° plane crosspolar lobes do not illuminate the perturbed portion of the profile.

6.2.6 *Overall crosspolar response*

Small aperture primary feeds suitable for illuminating the large opening angle of a conventional paraboloid frequently fall considerably short of the ideal in respect of crosspolar radiation.[7] This applies to types such as the open-ended waveguide, splash plate, dipole or dipole/reflector. More recently two special types of feed horn have been developed which yield primary pattern polarisation purity to the order of −35 dB to −40 dB. These are the 'ICR' choked circular waveguide, and the wide angle corrugated feed (see Chapter 5). Their primary pattern generated secondary crosspolar lobes are of the order of −45 dB.

Another significant source of crosspolarisation for small paraboloids is introduced by the feed support strut and/or feed transmission line aperture blockage. Fig. 6.10 illustrates the essential geometry of a small reflector for frequency re-use[2] from a communications satellite. The vertical 'strut' is actually a square section feeder waveguide for transmitting two orthogonally polarised signals to the feed horn. For the signal corresponding to polarisation parallel to the waveguide, off-axis crosspolarisation is introduced via two mechanisms:

(*a*) The electric element crosspolar pattern associated with induced currents in a strut (eqn. 8) is not well suppressed by the broad strut space factor pattern.

(*b*) Because of the thick waveguide/thin strut discontinuity in the vertical plane, current flows onto the horizontal struts at the central junction point.

At the certain frequencies where the square waveguide apparently behaves as a long resonant dipole, the strut generated crosspolar pattern is accentuated. Fig. 6.11 gives a measured crosspolar contour plot for these conditions. The horizontal plane symmetry together with vertical plane asymmetry are compatible with the support structure geometry.

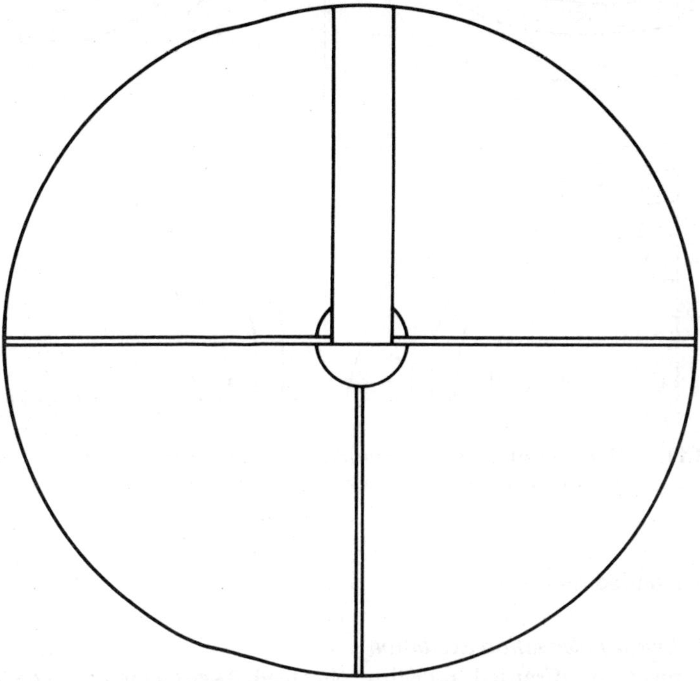

Fig. 6.10. *Effective aperture blockage geometry for a front-fed antenna*

In addition to struts, other less fundamental mechanisms often generate crosspolarisation in practical reflector antennas. These include orthogonal mode coupling in square or circular waveguide components, imperfections in orthomode transducers or circular polarisers, and mechanical misalignment of polarisation reference planes in the feed horn and/or feeder components.

Fig. 6.12 illustrates the measured overall crosspolar response for a small (28 λ) paraboloid designed for optimum polarisation properties.[2] The -36 dB worse case result may be roughly accounted for as the superposition of −45 dB lobes translated from the primary pattern, −46 dB lobes due to aperture blockage, and −50 dB lobes owing to feeder component imperfections.[8] The distinguishing feature of the latter is a finite crosspolar response on boresight.

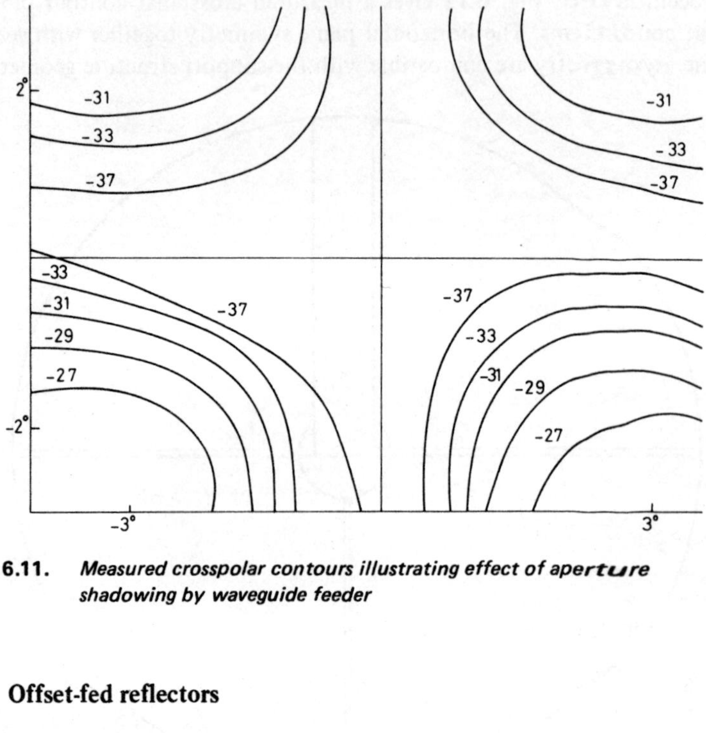

Fig. 6.11. *Measured crosspolar contours illustrating effect of aperture shadowing by waveguide feeder*

6.3 Offset-fed reflectors

6.3.1 Linear polarisation excitation

In a practical offset-fed reflector, the feed boresight axis $O_{z'}$ will ordinarily be tilted away from the axis of revolution O_z of the reflector generating surface to obtain an efficient illumination of the aperture. This contravenes the conditions stipulated in Section 6.2.2., and the transverse PO currents are no longer parallel.

Take

$$\angle zOz' = \theta_0$$

and suppose that the feed is copolarised *relative to its own $r'\theta'\phi'$ co-ordinate system.* Thus, from Section 6.1.1., the E-vector radiated by the feed makes an angle ϕ' with the $\mathbf{a}_{\theta'}$ unit vector. However, according to eqns. 18–24, to achieve x-directed transverse currents the ideal E-vector would be inclined at an angle ϕ to the $\mathbf{a}_\theta$ unit vector of

Fig. 6.13. But the conformal property of Section 6.2.2. specifies that any in-sphere misalignment of E translates simply to an equal in-plane misalignment of $\boldsymbol{i}$. Thus the current misalignment ψ_{tilt} is given by

$$\psi_{tilt} = \phi' - \phi + \psi = \tan^{-1} i_y/i_x \tag{6.33}$$

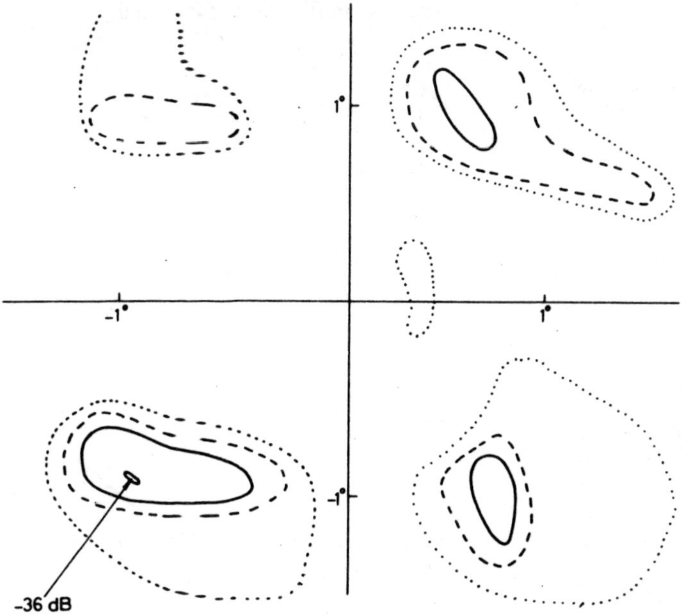

Fig. 6.12. *Measured crosspolar contours for a 28 λ front-fed reflector*

——— — 38 dB
- - - - — 40 dB
· · · · · — 42 dB

where ψ is the angle Q′PQ of the 'polarisation spherical triangle' illustrated in Fig. 6.13. Several important conclusions follow:

(i) The crosspolar to copolar induced current ratio is just the amount in radians by which the sum of the angles of the polarisation triangle exceeds 180°.

(ii) By a fundamental spherical trigonometry theorem, this is also the area of the triangle for unit sphere radius.

(iii) The crosspolar induced currents are thus approximately proportional to $\theta_0\theta'$, e.g. feed tilt × reflector opening angle.

(iv) The transverse induced currents take on the general type of distribution suggested in Fig. 6.14.

The area of a plane triangle with two specified sides maximises when these subtend a right-angle. It follows that i_y/i_x will nearly maximise for $\phi' = 0$. Application of the spherical cotangent formulas to this case yields

$$\tan \psi_{tilt} = \frac{i_y}{i_x} = \frac{1 - \cos \theta_0 \cos \theta'}{\cot \theta_0 \sin \theta' + \cot \theta' \sin \theta_0} \tag{6.34}$$

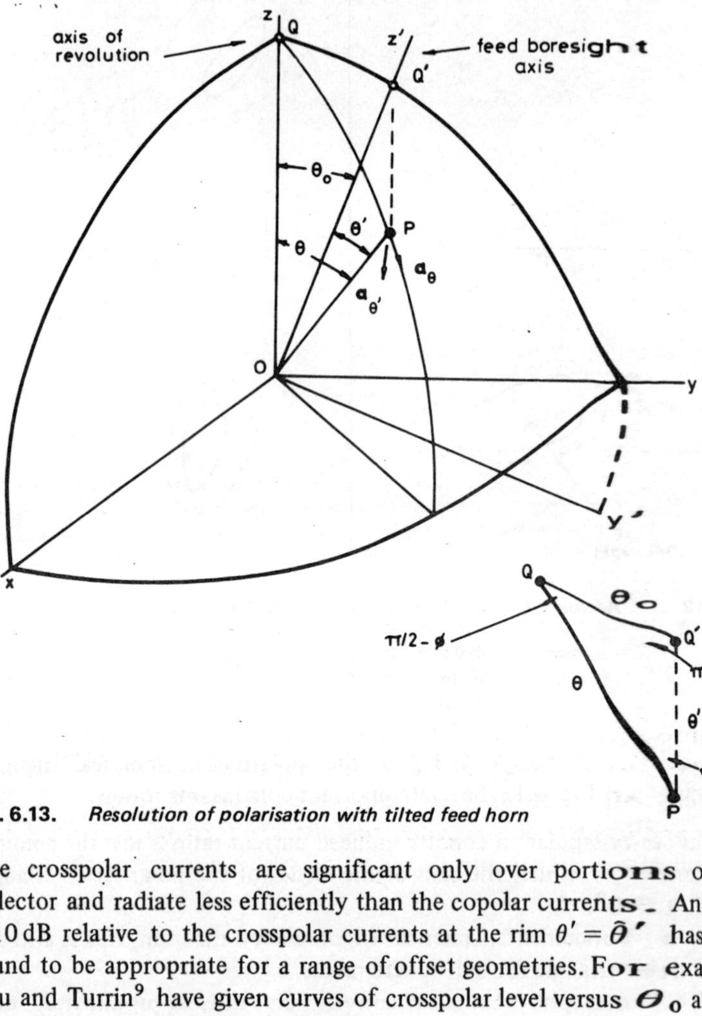

Fig. 6.13. *Resolution of polarisation with tilted feed horn*

The crosspolar currents are significant only over portions of the reflector and radiate less efficiently than the copolar currents. An η_c of -10 dB relative to the crosspolar currents at the rim $\theta' = \hat{\theta}'$ has been found to be appropriate for a range of offset geometries. For example, Chu and Turrin[9] have given curves of crosspolar level versus θ_0 and $\hat{\theta}'$. Their results are virtually duplicated by eqn. 34 x η_c. There is a 1 dB discrepancy for the somewhat unrealistic extreme case of a 90° offset feed and 90° reflector opening angle.

for RHCP and LHCP. According to eqn. 34, the phase shift will vary across a $\phi' = 0$ diameter of the reflector surface as an approximately linear function of θ'.

The rate of change of phase at the centre of the reflector is

$$\frac{\partial \psi_{tilt}}{\partial x} = \left|\frac{\partial \psi_{tilt}}{\partial \theta'}\right|_{\substack{\theta'=0\\ \phi'=0}} \left|\frac{\partial \theta'}{\partial x}\right|_{\substack{\theta'=0\\ \phi'=0}}$$

whence from eqn. 34,

$$\frac{\partial \psi_{tilt}}{\partial x} = (\operatorname{cosec} \theta_0 - \cot \theta_0)/r_0 \tag{6.35}$$

where

$$r_0 = 2F/(1 + \cos \theta_0) \tag{6.36}$$

represents the linear displacement of the $\theta' = 0$ point on the reflector from the focus.

This phase slope can readily be translated into a secondary pattern beam squint $\Delta\Theta$, resulting in the simple formula

$$\sin \Delta\Theta = \frac{\lambda}{4\pi F} \sin \theta_0 \tag{6.37}$$

suggested by Adatia and Rudge.[13] Approximation enters in eqn. 37 only in so far as the upper and lower parts of the reflector exhibit a slightly different phase slope to that for $\phi' = 0$.

References 6 and 14 describe an offset fed antenna for illuminating an elliptical region of the earth's surface from a communications satellite. This antenna employs frequency re-use via two hands of circular polarisation. The predicted CP contour plot is illustrated in Fig. 6.17*a* together with experimentally derived data for comparison. Fig. 6.17*b* compares contour plots for RHCP and LHCP, and clearly demonstrates the 'CP beam squint effect' discussed in previous paragraphs.

6.3.3 Depolarisation in circularly polarised antenna systems

As indicated in the previous Section, to the order of physical optics, the transverse induced currents in an offset reflector excited by a CP feed are themselves circularly polarised. However, as for the linear case of Section 6.2.4., both the vector element pattern (eqn. 10) and the axial currents are sources of residual depolarisation.

CP depolarisation effects are best understood in terms of the linearly polarised constituent excitations of the CP feed. Axial currents are excited with greatest amplitude in response to linear polarisation nominally parallel to the plane of offset. The axial current space factor then has a dominant 'even' component, as opposed to the 'odd' or monopulse variation of Section 6.2.4. In fact, it may be very approximately equated to the copolar transverse current space factor, yielding the result

$$P_x(\theta, \phi) \approx P_{co}(\theta, \phi) \tan \bar{\theta}_{norm} \sin \theta \cos \phi \tag{6.38}$$

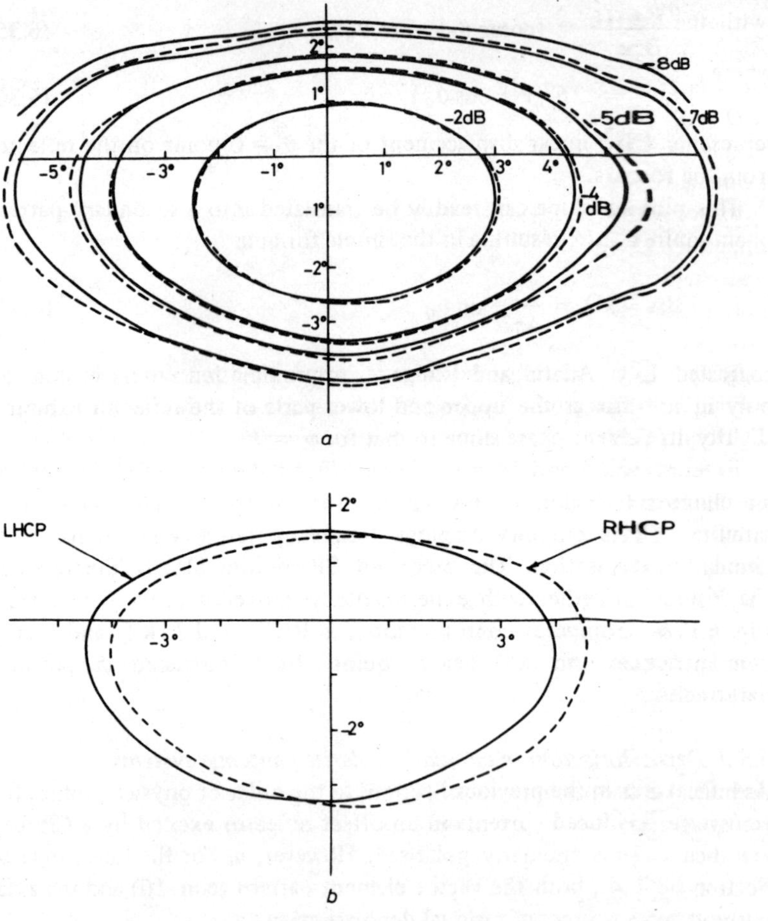

Fig. 6.17. *Contours for an offset fed elliptical beam antenna*
a Copolar measured (———) and predicted (– – – –)
b Measured contours illustrating CP beam squint

Here $\bar{\theta}_{norm}$ is the angle of the tangent to the offset surface at the centre of the aperture. This simple explicit formula is usually adequate to predict the level of depolarisation within about 2 dB.

Eqn. 38 specifies a crosspolar pattern with a null on axis and two lobes maximising in the symmetry plane, similar to the GO depolarisation case of Section 6.3.1. However, the crosspolar lobes are essentially in phase (or antiphase) with the copolar pattern, as opposed to phase quadrature for Section 6.3.1. Thus for a linearly polarised feed nominally aligned with the plane of offset, diffraction depolarisation is evidenced as a phase shift of the GO crosspolar lobe. For alignment with the symmetry plane, there is no parallel to this phenomenon. In consequence, there is net depolarisation for CP. As is suggested by the $\sin\theta$ factor in eqn. 38, this depolarisation vanishes in the optical limit $\lambda/D \rightarrow 0$.

Fig. 6.18 gives the unwanted hand contour plot for the elliptical beam antenna of Fig. 6.17. The PO/KHS analytical treatment takes into account diffraction depolarisation phenomena together with the imperfect polarisation performance of the primary feed horn. The latter was characterised in terms of measured E- and H-plane amplitude/phase pattern data for linear polarisation excitation, and the -35 dB feed crosspolar patterns are subject to a secondary pattern conversion efficiency of ≈ 1 dB as described in Section 6.2.5. The calculation is carried out assuming that pure CP is excited at the throat of the horn.

Despite the latter idealisation, the prediction is meaningful in the context of locating the 'worst case crosspolar' quadrant, corresponding to the constructive addition of feed horn and diffraction depolarisation unwanted hand generation mechanisms. The predicted level of the 'worst case' lobe is -34 dB, as opposed to a -31 dB measured value. At lower contour levels a more marked discrepancy between theory and measurement may be observed. Much of this is probably directly due to the 0·2 dB axial ratio of the polariser used for the real antenna (equivalent to -39 dB unwanted signal).

6.4 Dual reflector antennas

The scatter pattern of a hyperboloid subreflector can exhibit substantial diffraction depolarisation as described in Section 6.2.4. However, the crosspolar excitation efficiency η_c for the paraboloid main reflector is generally of the order of -5 dB. Thus the depolarisation lobes are to some extent depressed in the secondary pattern. Fig. 6.19 illustrates this feature for a 33 λ main reflector/5λ subreflector Cassegrain. For high efficiency shaped Cassegrainian systems (Chapter 7) the level and rate of change of illumination near the rim of the subreflector are ordinarily smaller than for a hyperboloid. It is found that depolarisation effects are always at a very low level in the secondary pattern. An example is the 621λ shaped Cassegrain described in Section 5.2.3, for which −63 dB secondary depolarisation is excited.

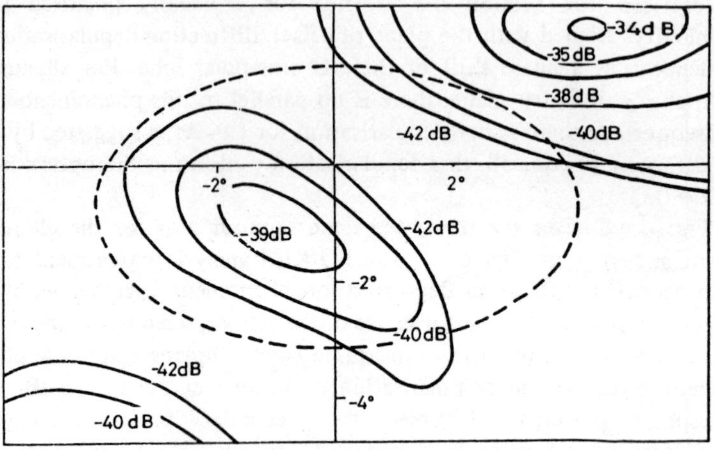

a

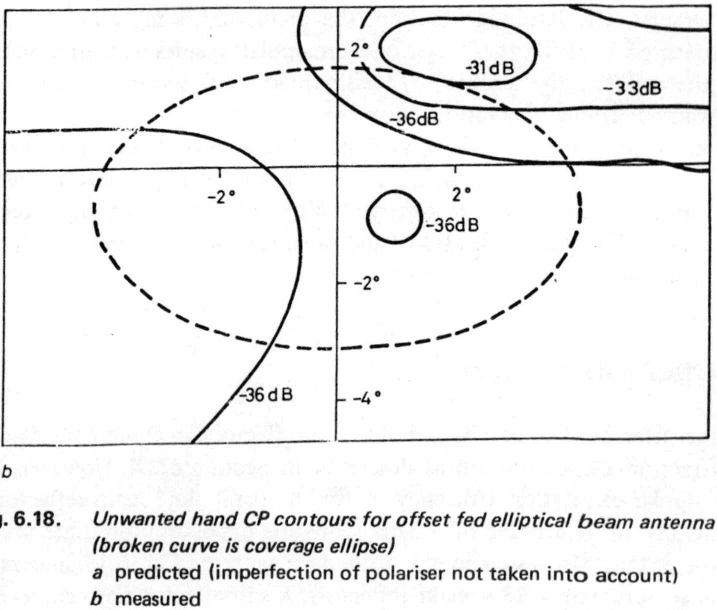

b

Fig. 6.18. *Unwanted hand CP contours for offset fed elliptical beam antenna (broken curve is coverage ellipse)*
a predicted (imperfection of polariser not taken into account)
b measured

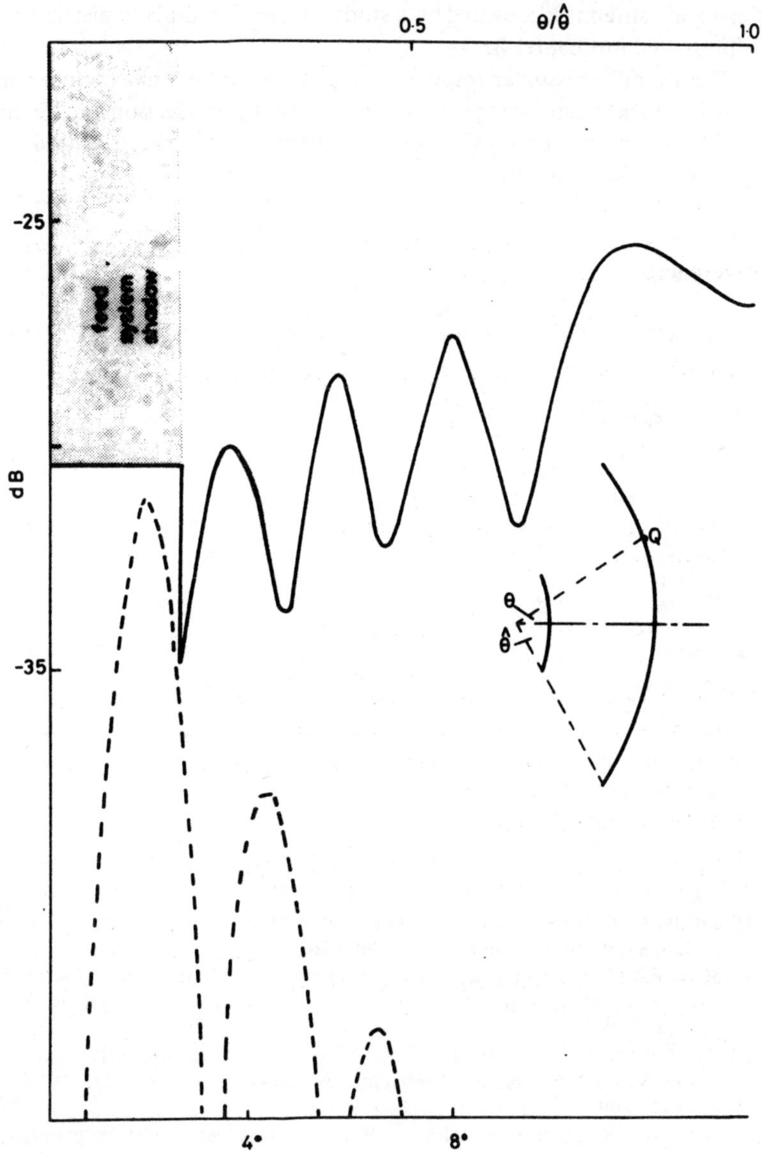

Fig. 6.19. *Illustrating suppression of subreflector depolarisation by main reflector*

——— relative subreflector scatter pattern crosspolar response at Q, $\theta/\hat{\theta}$ scale

– – – – relative secondary pattern crosspolar response

Further insight into diffraction depolarisation mechanisms in a Cassegrain antenna is provided by a study of the focal plane distribution, as presented in Chapter 8.

The overall crosspolar response of a Cassegrain antenna includes also contributions from the primary feed horn (see Sections 6.2.5 and Fig. 6.7) and the subreflector support struts, as for the front-fed case discussed in Section 6.2.6.

References

1 LUDWIG, A. C.: 'The definition of cross-polarisation', *IEEE Trans.*, 1973, **AP-21**, pp. 116–119

2 COWAN, J. H., and WOOD, P. J.: 'A shaped beam dual polarisation antenna for a regional satellite communication system'. European Microwave Conference (Brussels) 1973, **2**, C.4.3

3 WOOD, P. J.: 'Depolarisation with Cassegrainian and front-fed reflectors', *Electron. Lett.*, 1973, **9**, pp. 181–183

4 COWAN, J. H.: 'A shaped beam antenna for a maritime communications satellite'. IEE Antennas for Aircraft and Spacecraft Conference, London, 1975, pp. 101–106

5 WOOD, P. J.: 'Polarisation characteristics of reflector antennas' *Radio & Electron. Eng.*, 1975, **45**, pp. 750–852

6 WOOD, P. J. and BOSWELL, A. G. P.: 'Elliptical beam antenna for satellite applications'. IEE Antennas for Aircraft and Spacecraft Conference, London, 1975, pp. 83–88

7 RUDGE, A. W., et al.: 'Cross-polarised radiation from satellite reflector antennas'. Proceedings AGARD Conference on Antennas for Avionics, Munich, 1973, pp. 16.1–16.8

8 WOOD, P. J.: 'Crosspolarisation with Cassegrainian and front-fed reflectors', *Electron. Lett.*, 1973, **9**, pp. 597–598

9 CHU, T., and TURRIN, R. H.: 'Depolarisation properties of offset reflectors', *IEEE Trans.*, 1973, **AP-21**, pp. 339–345

10 CHEN, C. C., et al.: 'Crosspolarisation of offset paraboloid strip reflectors', P. Gap. Symposium, June 1974, pp. 260–261

11 SONDER, O.: 'Theoretical and experimental investigation of a double grid dual polarised antenna system'. U.R.S.I. Conference Proceedings, 1974, pp. 106–108

12 RUDGE, A. W., and ADATIA, N. A.: 'New class of primary feed antennas for use with offset fed parabolic reflector antennas', *Electron. Lett.*, 1975, **11**, pp. 597–599

13 ADATIA, N. A., and RUDGE, A. W.: 'Beam squint in circularly polarised offset fed reflector antennas', *ibid.*, 1975, **11**, pp. 513–515

14 COWAN, J. H., and ASHTON, R. W.: 'Elliptical beam antenna for satellite applications, Part 2 design and development'. IEE Antennas for Aircraft and Spacecraft Conference, London, 1975, pp. 89–94

Chapter 7

Shaped Cassegrain design

7.1 Cassegrainian reflector systems using geometrical optics profiles

7.1.1 Concepts

One of the most interesting aspects of reflector antenna design is the possibility of exploiting the geometrical degrees of freedom inherent in the profile of a reflector in order to achieve special electrical properties. The necessary capability, both for numerical analysis and synthesis, and for manufacture of 'shaped' reflecting surfaces, has become readily available in recent years. For example the manufacturing problem can be resolved by machining a master or mould under control of data from a paper tape. In consequence, applications involving profiles which are intentionally distorted relative to classical conic surfaces are becoming increasingly common.

In Cassegrain feeds, there has been a major incentive to improve the overall gain and efficiency relative to the traditional hyperbola/parabola antenna. For the latter configuration, an upper bound to η_t is set by the necessary compromise between uniformity of illumination of the final aperture, and the proportion of energy radiated by the feed horn that is 'captured' by the subreflector.

Galindo[1] was the first to demonstrate that the hyperbola/parabola is only one member of an infinite set of Cassegrain profile combinations having two optical foci. That is, there are an infinity of solutions which allow the parallel rays associated with an axially incident plane wave to be brought to a point focus. The latter becomes the secondary focus of the Cassegrainian system, and conceptually at least coincides with the centre of the feed aperture. However, a true primary focus (corresponding to the confocal point of the parabola and hyperbola of a traditional Cassegrain antenna) is no longer in evidence.

In Fig. 7.1, the surface of the subreflector has been perturbed relative to a basic conic section geometry in order to redistribute the energy of the primary feed in favour of the wider angle GO rays. This redistribution is suggested by the ray concentration near the rim of the main reflector.

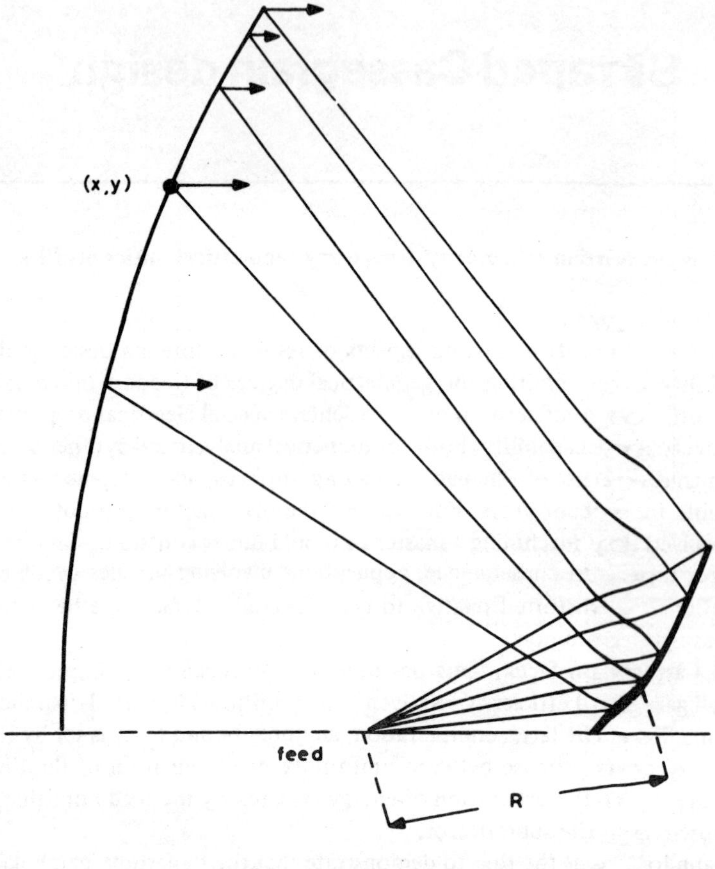

Fig. 7.1. *Optical ray paths in Williams Cassegrain*

If such a modification were applied to the Cassegrainian feed for a parabolic reflector, the collimating action of the latter would inevitably be degraded. Thus the total GO path length from the secondary focus to an arbitrary point on the final aperture would not be a constant quantity. However, following the approach proposed by Galindo, it becomes possible to restore collimation by introducing an appropriate small distortion of the main reflector.

Williams[2] proposed an implementation of these principles in which the aim was to synthesise a uniform GO illumination at the main reflector surface. For a specified primary feed horn pattern, the required profiles are derived by solving numerically a system of simultaneous differential equations, the unknowns being R, x, and y (see Fig. 7.1).

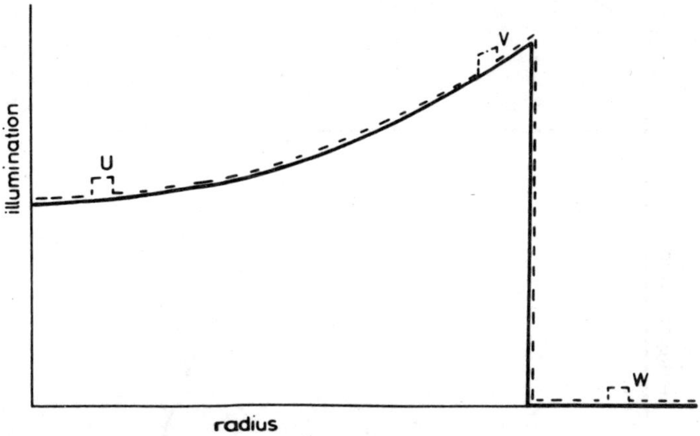

Fig. 7.2. *Ideal Cassegrainian feed system pattern specified by geometrical optics principles*

The GO subreflector scatter pattern for a Williams Cassegrain has a sector limited shape (Fig. 7.2)

$$P(\theta) \approx \sec^2 \theta/2 \qquad \theta \leqslant \theta_B \tag{7.1}$$

The RHS of eqn. 1 represents the law which is appropriate to provide compensation for the space attenuation for a paraboloid. However, when the scatter pattern is calculated by a vector diffraction method (for example, as for Fig. 5.14) other features become evident. In particular it is found that a finite pattern slope persists into the GO shadow region, $\theta > \theta_B$. This implies that a significant portion of the energy in the subreflector scatter pattern will spill past the main reflector.

7.1.2. Modified Williams antenna

For the modified Williams antenna, the aim is to synthesise a GO aperture distribution similar to Fig. 7.3, rather than a uniform illumination. The optimum form and extent of the aperture taper for a particular Cassegrain geometry has to be established by trial and error. Thus, vector diffraction calculations are repeated until an acceptable compromise between uniformity of illumination and spill-over at the main reflector is found. The improvement relative to the basic Williams antenna is generally small, perhaps 0·1 dB in gain. Nevertheless, it can be worthwhile for large antennas where the cost of obtaining an extra dB of gain by increasing the overall size of the antenna is considerable.

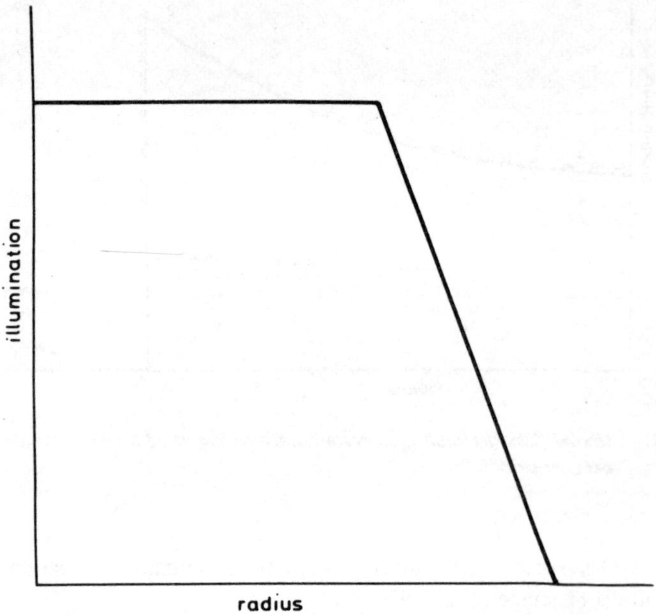

Fig. 7.3. *Ideal aperture illumination specified for modified Williams Cassegrain antenna*

The reflector profiles for the modified Williams antenna are conveniently represented as perturbations relative to 'base' conic sections (see Section 7.2.2). Fig. 7.4 illustrates the perturbation curves for a typical modified Williams configuration.

7.1.3 Feed refocusing

One of the deficiencies of GO profile synthesis is that it cannot take into account near field perturbations. Thus, usually the subreflector will not be sufficiently distant from the feed horn to be illuminated by a true far field pattern. A similar consideration applies in respect of the main reflector, with the subreflector in the role of a primary source.

Fig. 7.5 gives the geometry of a 275 λ diameter Williams Cassegrain. Vector diffraction calculations based on the S.W.E. techniques of Chapter 3 demonstrate the presence of significant phase errors due to near field effects. That is, the phase of the subreflector scatter pattern, as evaluated at finite range at the surface of the main reflector, deviates from the ideal phase that would be predicted by GO ray tracing. This is indicated by curve C of Fig. 7.6.

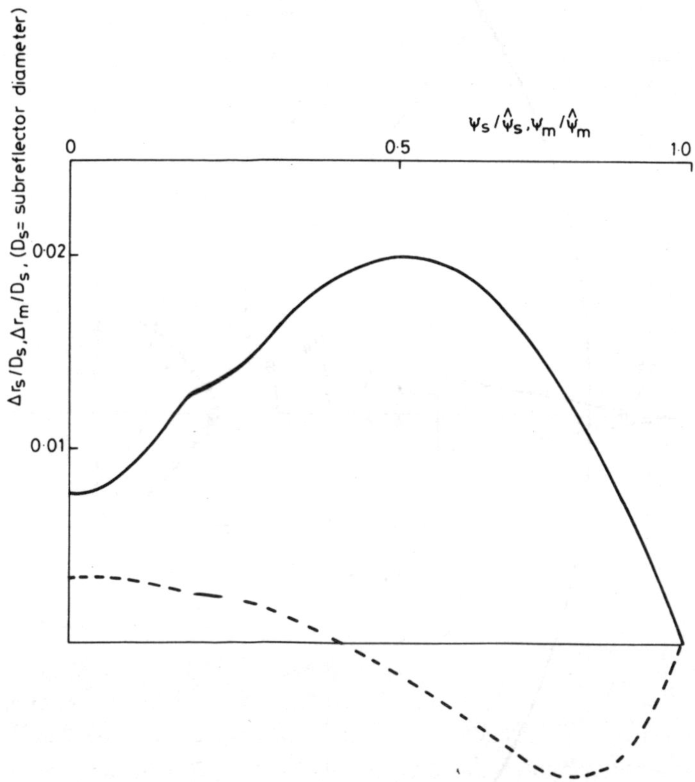

Fig. 7.4. *Normalised perturbation profiles for a Williams antenna*
——— subreflector
- - - - main reflector

The remaining two curves of Fig. 7.6 illustrate the phase errors present in the secondary focus focal plane. The receiving field in this plane (curve A) is calculated by evaluating the successive scattering of an axially incident plane wave by the main and subreflectors (see

Chapter 8). The transmitting field in the same plane amounts to the feed horn aperture field. This also varies in phase (curve B) owing to wavefront curvature arising from a finite flare angle.

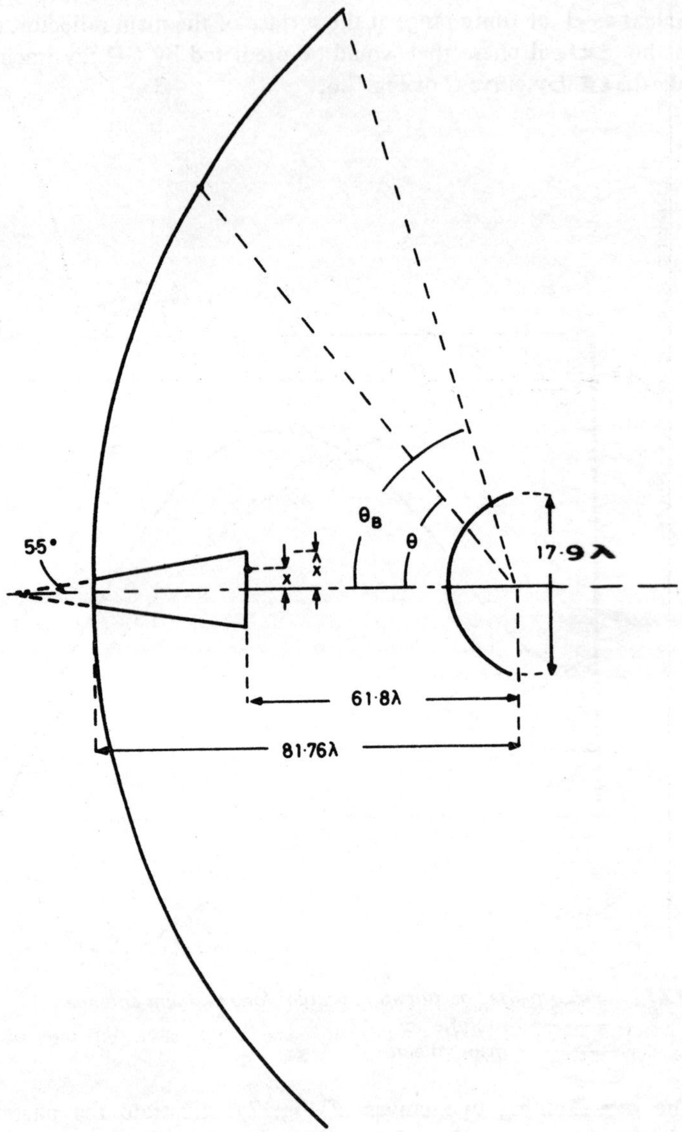

Fig. 7.5. *275 λ modified Williams Cassegrain geometry* ($\theta_B = 80{\cdot}12°$; $\hat{x} = 4{\cdot}945\ \lambda$)

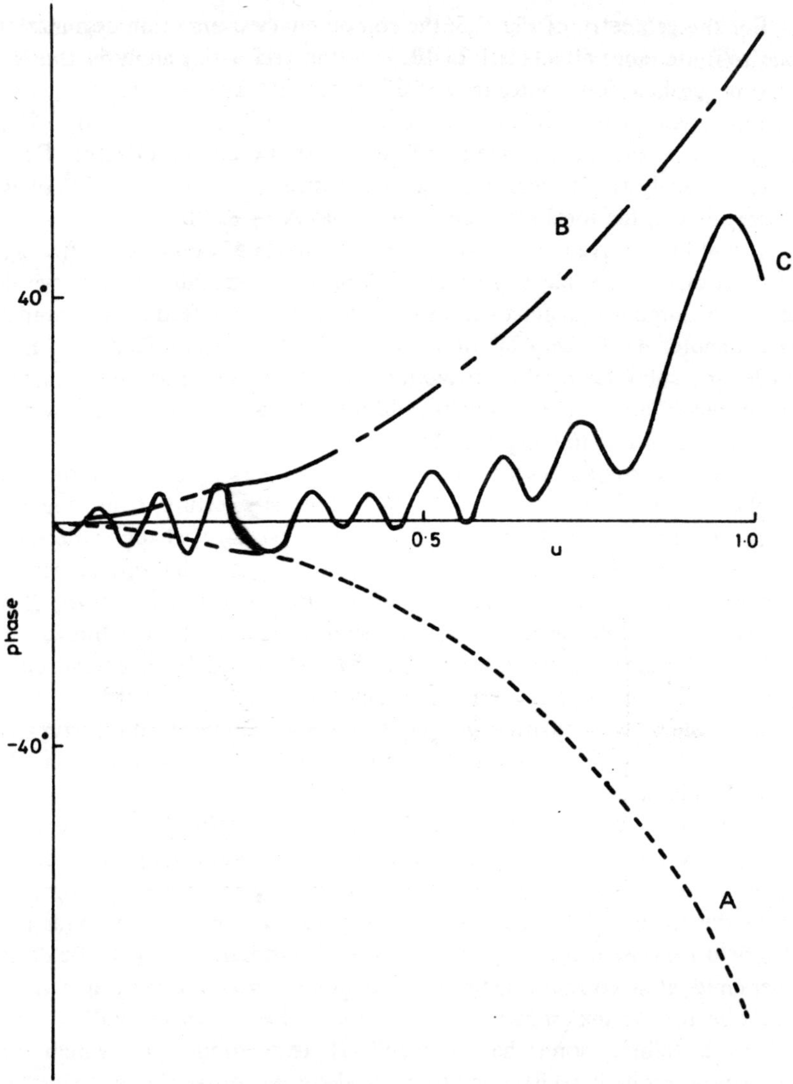

Fig. 7.6. *Phase errors in Williams Cassegrain*
---- phase advance in focal plane as a function of $u = x/\hat{x}$ (receiving)
— — — phase delay at primary feed aperture as a function of $u = x/\hat{x}$ (transmitting)
——— phase advance relative to a forward propagating plane wave at main reflector suface as a function of $u = (\theta - \theta_A)/(\theta_B - \theta_A)$ (transmitting)

For the geometry of Fig. 7.5, the computed overall gain degradation due to finite range effects is 1·25 dB. In a transmitting analysis, this loss may be broken down into an 0·55 dB contribution due to the scatter pattern phase errors evident from curve C of Fig. 7.6, and an 0·7 dB contribution due to increased spill-over past the subreflector. For a receiving analysis, the loss arises almost entirely from the 150° phase discrepancy in the focal plane between curves A and B.

It will be observed that curves A and B are of opposite relative sign in relation to the ideal 'complex conjugate' condition that would ensure maximum transfer of received power to the feed horn terminals (see Chapter 4). In general, finite range effects, finite feed horn flare angle, and subreflector rim diffraction each introduce phase errors, and these phase errors are mutually additive in a Cassegrainian system designed according to GO principles.

A remedy sometimes adopted is to 're-focus' the feed horn. This implies an axial displacement of the horn aperture from the GO secondary focus towards the subreflector. In a practical application, the extent of this displacement would ordinarily be determined via an experimental phase centre measurement for the feed horn. This measurement could be carried out in various different ways, but for the present purposes it is useful instead to adopt an analytical simulation of the measurement process. For the geometry of Fig. 7.5, this has been accomplished by calculating the phase characteristics of the corrugated conical feed horn by using the Laurent S.W.E. theoretical model described in Section 5.1.

The vertical bar to the left of Fig. 7.7 identifies the nominal GO focus position for the feed aperture. The remaining bars define alternative refocused positions for which a phase centre coincides with the focus point. The first type of phase centre corresponds to a far field measurement situation. The second assumes that the fields are measured at a constant range corresponding to the feed aperture – subreflector vertex separation. The final bar corresponds to the ultimate (albeit somewhat impractical) technique in which the measuring probe is constrained to move along a curve similar to a radial cut of the subreflector. It is assumed that the feed horn is simultaneously displaced axially until the best available approximation to the phase characteristics of a point source placed at the focus position is obtained.

The solid curve in Fig. 7.7 indicates that an η_t of 0·868 can be realised for the 275 λ modified Williams Cassegrain. To achieve this optimum solution requires more refocusing than predicted by phase centre measurements. Moreover, the resulting geometry has undesirable features in that the feed horn is brought well forward from the main reflector and will need mechanical support and a waveguide run.

It might be imagined that it would be advantageous to start with a geometry which puts the subreflector more into the far field of the feed horn. This might be accomplished either by using a smaller feed horn closer to the subreflector, or by using a larger subreflector. However, the geometry of Fig. 7.5 has refocused into near MPE (minimum phase error, see Section 5.1.3) properties for the feed horn. In fact, it is just for this reason that refocusing has proved so effective. An alternative initial geometry might need less refocusing but would not be more efficient.

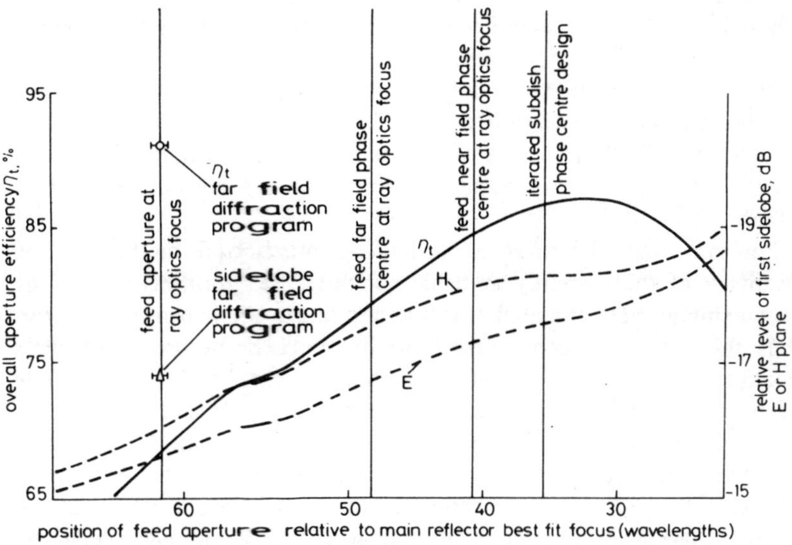

Fig. 7.7. *Performance of ray optics shaped reflectors with feed focusing* (not including aperture blockage by feed horn)

7.2 Diffraction profile synthesis

7.2.1 Specification of optimisation criteria

In its most basic form, the GO synthesis technique aims to synthesise the sector scatter pattern of eqn. 1 and Fig. 7.2. Undeniably, such a pattern would provide an ideal uniform illumination of the final aperture. In practice, however, a subreflector of finite diameter can only approximate Fig. 7.2. Thus, it is necessary to stipulate *not only the ideal pattern, but the relative importance of different types of deviation from it.*

For example, when viewed in terms of overall antenna performance, the perturbations *U*, *V*, *W* of Fig. 7.2 have quite different effects. Equally, within the main reflector capture angle, a phase quadrature perturbation is considerably more deleterious than a cophasal perturbation. Outside the capture angle however, either is of the same importance.

So long as the concept of the synthesis method remains unchanged, e.g. subreflector profile varied to approximate an idealised scatter pattern, it is extremely difficult to formulate universal quantitative optimisation rules. In later sections of this chapter it will become apparent just how complex such criteria would have to be. In GO synthesis the effect of the limited number of degrees of freedom imposed by the finite subreflector is much the same as truncating a Fourier series. That is, vector diffraction calculations for the synthesised profile demonstrate that something approaching a 'least-squares fit' to the idealised scatter pattern is generated. Later synthesis techniques by Daveau[3] and Poulton[4] use a vector diffraction treatment from the outset, but in practice inevitably become a 'least squares' method by default.

Poulton[5] has described a further optimisation method taking advantage of the currently available computer algorithms for finding the maximum value of a nonlinear function of an *N*-dimensional vector. Thus the individual 'components' of the vector become parameters specifying the reflector profile. In one version of Poulton's technique the function itself is identified with antenna gain. It is equally possible to formulate the latter in terms of a current integration at the main reflector surface or in terms of a correlation integration at the subreflector, as described in Chapter 4. Moreover, the method has an inherent degree of flexibility in that it can be arranged that the function gives a weighting to additional features of importance in the optimisation process. In the implementation described in Reference 5, only the subreflector is shaped, assuming a simple 'point source' feed model and that the far field criterion may be applied. The deficiencies that can arise as a result of the latter assumption have already been cited in Section 7.1.3. Potentially, however, Poulton's technique appears to offer a direct means of obtaining an optimum compromise between gain and sidelobe performance in a shaped Cassegrain system, a result which can be difficult to achieve by any other route.

The purpose of this Section is to introduce a fundamentally different approach. In the DPS technique[6–8] the profile of a reflector is perturbed whilst aiming to conjugate match two fields *at its surface,* rather than to generate a specified distant field distribution. The two

(*b*) An initial estimate of the best fit phase centres for the transmit and receive patterns is made. Preferably the position of the transmit centre C_1 will be derived either by analytical modelling (e.g. see Section 5.1.3), or from measured feed horn phase pattern data, or from an S.W.E. fit to measured data (see Section 5.2.3). Alternatively, C_1 may be assigned 'by default', for example as a point coincident with the centre of the feed horn aperture. The penalty for this latter choice will ordinarily be a slower convergence of iterations.

The receive phase centre C_2 is usually taken to be coincident with the base paraboloid focus. The centres C_1 and C_2 correspond to the $r'\theta'\phi'$ and $r\theta\phi$ spherical polar co-ordinate systems to be used in the analysis. The focal length of the base hyperboloid will be chosen to correspond to the displacement C_1C_2 when the feed horn is placed in its desired position.

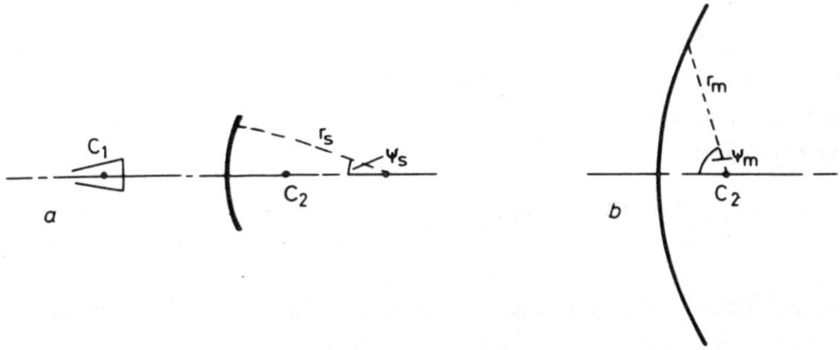

Fig. 7.8. *Co-ordinate systems for reflector profiles*
a subreflector
b main reflector

(*c*) The 'transmit' field incident on the subreflector is specified as a Laurent S.W.E. for unit power input to the feed horn. Empirical and analytical methods for evaluating the S.W.E. weights have been given in Chapter 5. The series is re-expanded to determine the electric field component

$$E_t = f_1^t(\theta') \cos\phi' \boldsymbol{a}_{\theta'} - f_2^t(\theta') \sin\phi' \boldsymbol{a}_{\phi'} + f_3^t(\theta') \cos\phi' \boldsymbol{a}_{r'} \quad (7.2)$$

at finite range at the surface of the subreflector. Modes of azimuthal dependence of higher order than $\cos\phi'$ are not included in the expansion, since they cannot influence the correlation integral of eqn. 4.5.

The 'receive' field at the subreflector

$$H_r = f_1^r(\theta) \sin\phi \boldsymbol{a}_\theta + f_2^r(\theta) \cos\phi \boldsymbol{a}_\phi + f_3^r(\theta) \sin\phi \boldsymbol{a}_r \quad (7.3)$$

is evaluated as a Taylor S.W.E. In this case the weights are specified by matching to the PO currents induced in the main reflector by an axially incident plane wave (see Section 3.3.4 and Reference 2 of Chapter 3).
(*d*) Average phase patterns at the subreflector surface

$$\Phi_t^{AV} = \angle\{f_1^t(\theta') + f_2^t(\theta')\} \tag{7.4}$$

and

$$\Phi_r^{AV} = \angle\{f_1^r(\theta) + f_2^r(\theta)\} \tag{7.5}$$

are now defined.
(*e*) To achieve a conjugate phase match at the subreflector as required by the field correlation theorem

$$\Phi_{rt} = \Phi_t^{AV} + \Phi_r^{AV} \tag{7.6}$$

should be independent of ψ_s. If in addition Φ_{rt} is to maintain its base profile value Φ_{rto} at the subreflector rim,

$$\Phi_{rt} = \Phi_{rto} \tag{7.7}$$

For the moment, assume the transmit and receive wavefronts are spheres concentric with C_1 and C_2, respectively. Then a profile perturbation

$$\Delta^u r_s = \frac{\Phi_{rt} - \Phi_{rto}}{2k} \sec(\theta' + \theta_{norm}) \sec(\psi_s - \theta_{norm}) \tag{7.8}$$

would ideally restore phase correlation. In fact, the DPS concept only requires that this perturbation should correspond to some improvement in η_t. Any approximation inherent in eqn. 8 will be eliminated when the iterations are continued to convergence.
(*f*) The profile perturbation as calculated in step *e* will tend to exhibit rapidly oscillatory 'diffraction ripples' superimposed on a smoothly varying curve. Averaging of the *E*- and *H*-plane patterns via eqns. 4 and 5 has already been introduced with the aim of reducing the ripples to a second-order effect. Nevertheless, the residual deviation from smoothness is sometimes sufficient to pose a manufacturing problem, as well as creating somewhat of a 'credibility gap' in relation to the use of the physical optics principle.

The superscript '*u*' on the LHS of eqn. 8 implies 'unsmoothed'. Profile perturbations are smoothed by applying a 'cosine profile filter' transformation. Thus, if

$$\gamma = \sin^{-1}(\psi_s/\hat{\psi}_s) \tag{7.9}$$

then for the range

$$0 \leqslant \psi_s \leqslant \hat{\psi}_s \tag{7.10}$$

the transformed perturbation $\Delta^u r_s(\gamma)$ is periodic, continuous, and representable as the discrete Fourier series:

$$\Delta r_s(\gamma) = \sum_{j=0}^{j=S_L} c_j \cos(2j\gamma) \tag{7.11}$$

where

$$c_j = \frac{2\delta(j)}{n-1} \sum_{i=1}^{i=n} q(i)\Delta_i^u r_s \cos\left(j\pi\left\{\frac{i-1}{n-1}\right\}\right) \tag{7.12}$$

with

$$\delta(j) = \begin{cases} \frac{1}{2} & \text{if} \quad j = 0 \\ 1 & \text{if} \quad j > 0 \end{cases} \tag{7.13}$$

and

$$q(i) = \begin{cases} 1 & \text{if} \quad i = 1 \\ 2 & \text{if} \quad 1 < i < n \\ 1 & \text{if} \quad i = n \end{cases} \tag{7.14}$$

In eqn. 12 it is assumed that $\Delta^u r_s$ is specified at n ordinates $\Delta_i^u r_s$ equally spaced in γ in the range 0 to $\pi/2$. These ordinates may be derived from the primary specification of the profile data (e.g. equally spaced intervals of ψ_s, see step *a*) by local second-order numerical interpolation.

If S_L is large, eqn. 11 tends to approximate the original unsmoothed perturbation. By assigning a small S_L, smoothing is effected. Typically,

$$S_L \approx 10 \tag{7.15}$$

gives rise to an adequate smoothing constraint whilst not introducing any significant degradation of the electrical performance.

The new smoothed subreflector profile is specified as

$$r_s^1(\psi_s) = r_s^b(\psi_s) + \Delta r_s^1(\psi_s) \tag{7.16}$$

Here the superscript '*b*' implies base profile, and the superscript '1', first iteration.

(*g*) Steps *a*–*f* are now repeated, but in the context of fields at the *main reflector* surface. The 'receive' fields are now simply those of an axially incident plane wave here. The 'transmit' fields, however, are obtained by evaluating successively the finite range fields of the feed horn at the surface of the subreflector (Laurent S.W.E., Chapter 5), and the scattered fields of the subreflector at the surface of the main reflector (Laurent S.W.E., Section 3.2.7.1).

The new, smoothed, main reflector profile is specified as

$$r_m^1(\psi_m) = r_m^b(\psi_m) + \Delta r_m^1(\psi_m) \tag{7.17}$$

(h) The existence of a new main reflector profile implies that the 'receive' fields *at the subreflector* will now have changed. Therefore the next step is to re-evaluate them, and to calculate a new perturbation to the current subreflector profile $r_s^1(\psi_s)$, as per steps c–f.
(i) An updated subreflector profile (now $r_s^2(\psi_s)$) implies that the 'transmit' field at the main reflector will have undergone some change. This field is re-evaluated, and step g repeated to update the main reflector profile to $r_m^2(\psi_m)$.
(j) Iterations are continued in this fashion. A correlation of fields across the subreflector according to eqn. 4.5 provides a convenient test to determine when the convergence point is reached. The final profiles may be written in the form

$$r_s^f(\psi_s) = r_s^b(\psi_s) + \Delta r_s^f(\psi_s) \tag{7.18}$$

and

$$r_m^f(\psi_m) = r_m^b(\psi_m) + \Delta r_m^f(\psi_m) \tag{7.19}$$

where

$$\Delta r_s^f = \Delta r_s^1 + \Delta r_s^2 + \ldots \tag{7.20}$$

and

$$\Delta r_m^f = \Delta r_m^1 + \Delta r_m^2 + \ldots \tag{7.21}$$

7.2.3 *An example*

We select as an example a Cassegrain having a 150 λ diameter main reflector and a 13·5 λ subreflector. Fig. 7.9 illustrates the initial transmit and receive fields at the surface of the base hyperboloid (step c above), together with the differential phase Φ_{rt} (step e above). The latter curve may be seen to exhibit diffraction ripples, the effect of which carries through to eqn. 8. The ripples are smoothed at each synthesis iteration according to eqns. 9–14. With moderate smoothing this process results in a slight 'bump' near the centre of the profile perturbation curve (see Fig. 7.10). This behaviour is much less noticeable in relation to the overall subreflector profile, however. If desired, the 'bump' can be eliminated completely whilst still retaining most of the advantage of DPS synthesis.

The synthesised profiles are found to virtually eliminate the phase error evidenced in Fig. 7.9. The amplitude correlation between transmit and receive fields has also been improved somewhat, as may be seen by comparing Fig. 7.11 to Fig. 7.9. The improvement would be more dramatic for a large Cassegrain, where shaping of the main reflector plays a more important role.

The final subreflector scatter pattern (broken curve of Fig. 7.12) has a somewhat irregular appearance in relation to a Williams-type design. For example there is a noticeable dip at ≈60° off-axis. In fact, the dip is a kind of 'Gibbs phenomenon', and is an essential prerequisite to achieving a steep pattern slope at the 80° edge angle.

It is interesting that this feature is traceable to the optimisation criteria only in a very indirect fashion (e.g., the criteria in respect of subdish shaping are based on fields at the subdish surface, and not on

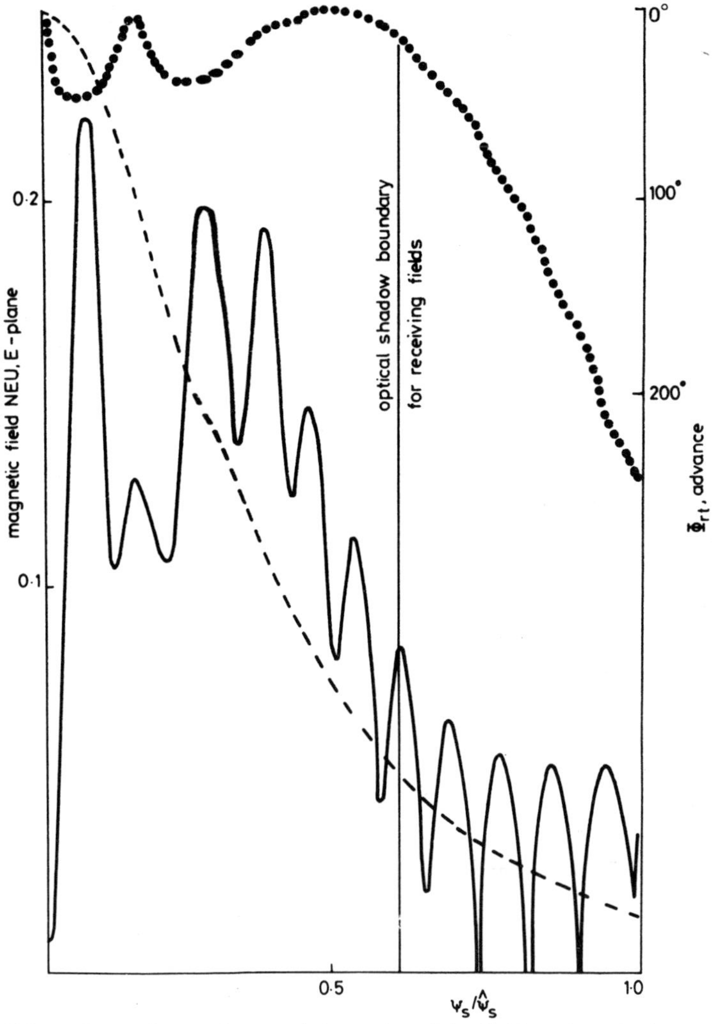

Fig. 7.9. *Comparison of transmitting and receiving fields at the subreflector, base profile case*

——— receiving
— — — transmitting
• • • differential phase Φ_{rt}

the scatter pattern). Nevertheless, the DPS synthesis process automatically ensures that the best possible compromise is achieved between nonuniformity of aperture distribution owing to the 60° dip, and spill-over past the main reflector owing to finite pattern slope. In

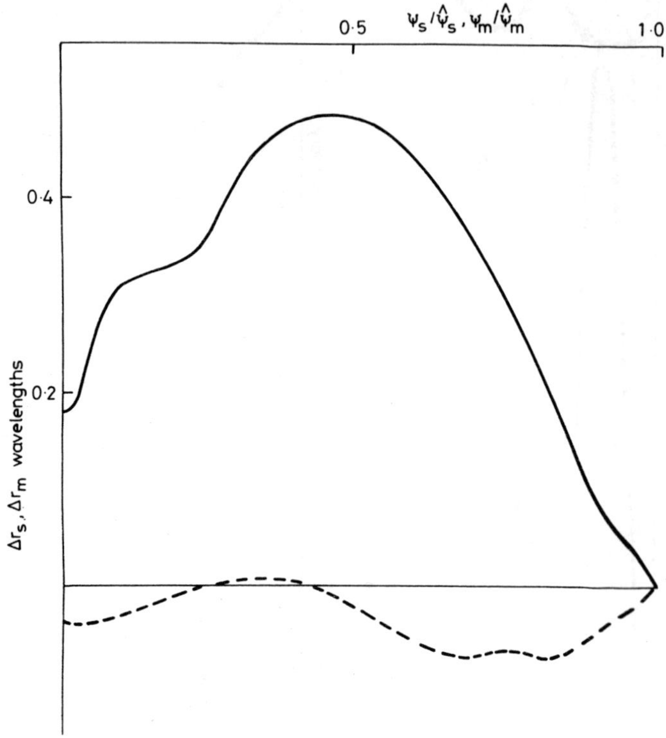

Fig. 7.10. *DPS perturbation profiles for a 150 λ Cassegrain*

——— subreflector
– – – – main reflector

fact, the slope for DPS reflectors is typically 1·75 times larger than a comparable ray optics profile design.

Another feature of the DPS scatter pattern arising indirectly from the synthesis process is the central null, as will now be described.

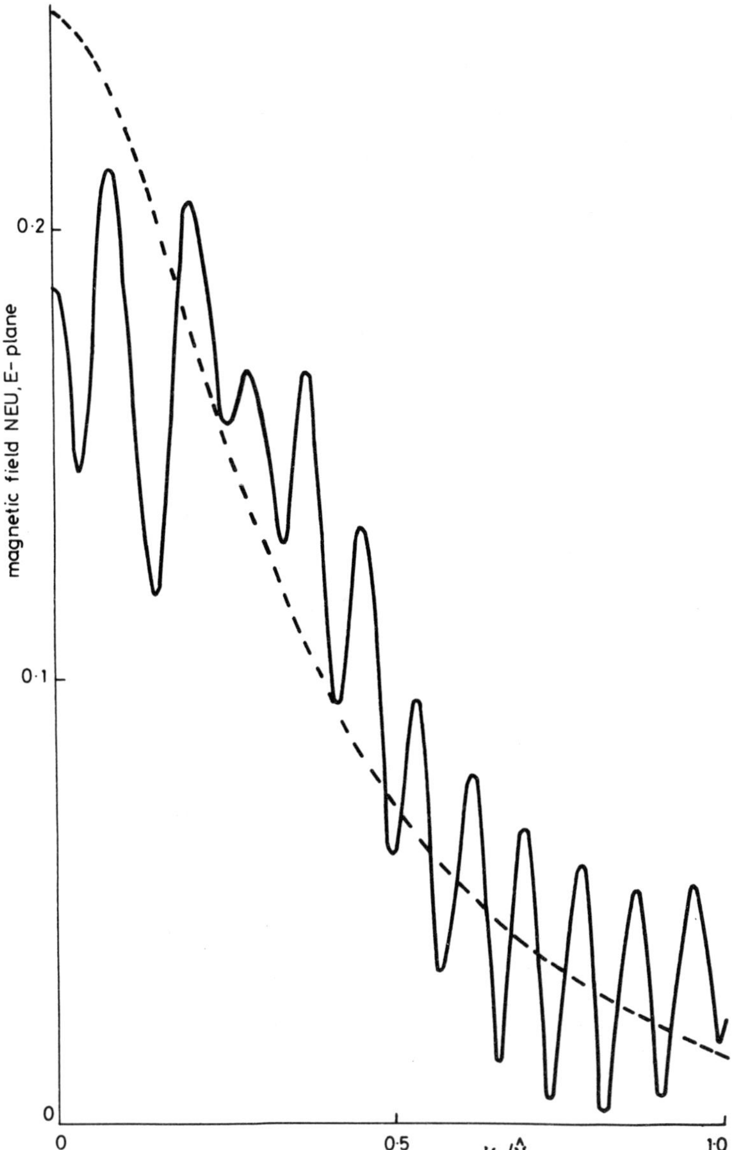

Fig. 7.11. *Comparison of transmitting and receiving fields at the subreflector, final synthesised profiles*

——— receiving

– – – – transmitting

7.3 Optimisation for central null in scatter pattern

As suggested by Fig. 7.13, to the order of GO the aperture of the main reflector will be shadowed at the centre owing to subreflector and feed aperture blockage effects. This shadowing effect is incorporated into the Taylor S.W.E. formulation of the receive fields incident on the subreflector by stipulating that the main reflector induced currents are zero in the shadow area.[9]

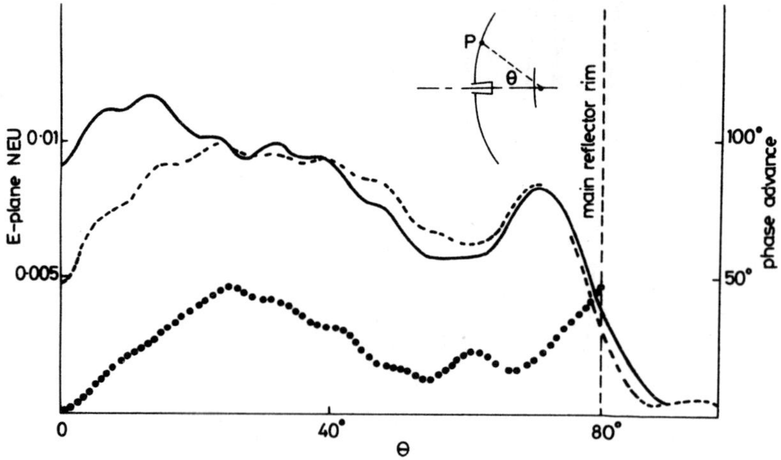

Fig. 7.12. *Fields at main reflector surface for DPS Cassegrain*
——— partially shaped
– – – – fully shaped
• • • partially shaped, phase relative plane wave

In the transmitting case, power in the region

$$\theta \leqslant \theta_A \tag{7.22}$$

is 'blocked', e.g. returned to the feed horn or scattered into wide angle radiation. The ideal scatter pattern would therefore have a null in this sector, to avoid wastage of power. However, the capabilities of a finite subreflector to generate such a null are limited. Moreover, if all the inherent degrees of freedom of the profile are to be used up in pursuit of this one aim, other beneficial features of the scatter pattern will be sacrificed.

It is for these reasons that the scatter pattern of Fig. 7.12 exhibits only a relatively shallow null. This represents the compromise between blockage effects and uniformity of illumination of the final aperture which gives optimum η_t for the particular 150 λ Cassegrain geometry involved. For other configurations the null may be deeper, or alternatively non-existent.

One part of the shadowing effect of Fig. 7.13 is still present in a feed system (feed and subreflector) pattern. Thus a measured feed system pattern will always show a central disturbance due to the feed

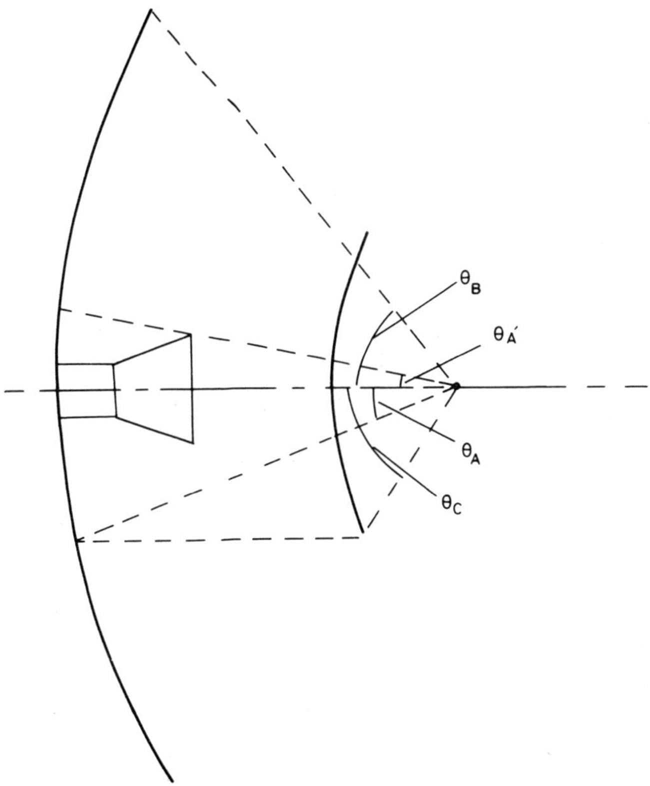

Fig. 7.13. *Geometrical parameters for Cassegrain central shadow*

horn shadow. With the usual Cassegrain configuration, this effect is difficult to distinguish experimentally from the inherent null of Fig. 7.12. The latter is of course present before the subreflector scattered fields reach the feed horn.

However, it is possible to hypothesize a special geometry for which $\theta_A \gg \theta_{A'}$, and the inherent null of the shaped subdish scatter pattern is much wider than the feed shadow. A feed system having a $\theta_A = 40°$,

$\theta_{A'} = 7{\cdot}5^\circ$ geometry was constructed to realise these conditions experimentally. The steep-sided central null of Fig. 7.14 is remarkable in that it is generated by a subreflector of essentially smooth profile of $20{\cdot}4\,\lambda$ diameter. This type of response could not be realised via ray optics profile synthesis of the Galindo type.

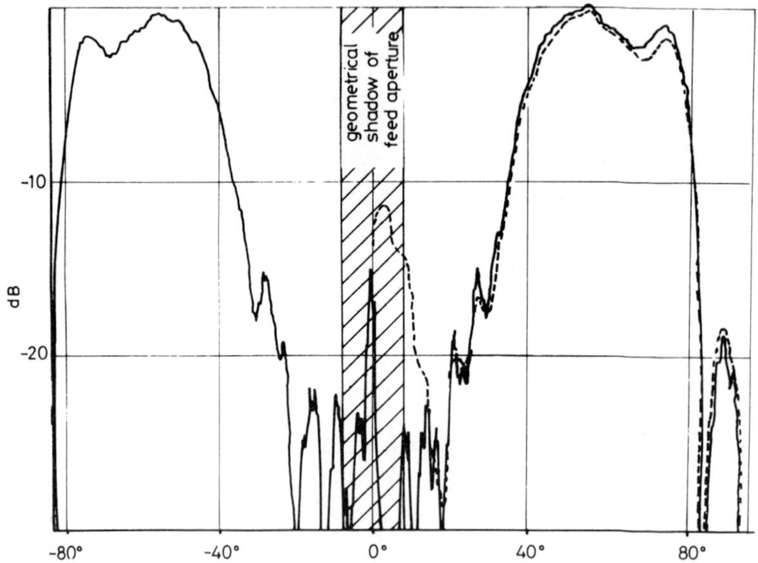

Fig. 7.14. *Scatter pattern for experimental DPS subreflector*
---- predicted
—— measured

7.4 Focal plane patterns

The beneficial properties of DPS profiles can also be recognised in the secondary focus focal plane patterns. For the $150\,\lambda$ Cassegrain, comparison of Fig. 7.15 with Fig. 7.6 indicates the improvement in phase correlation obtained. In terms of a focal plane analysis, η_t is limited more by spill-over than by imperfect phase or amplitude correlation within the feed aperture in the DPS case.

7.5 Some design criteria for DPS shaped Cassegrains

7.5.1 Selection of base profiles

The first rule to be applied in selecting base profiles is that they should form an optically focused system. In addition, there are further considerations to be taken into account relating to the diameter D_s of the subreflector. When the latter is relatively small in terms of wavelengths, the Williams-type solution differs markedly from the DPS solution. The standard confocal hyperbola/parabola combination (or confocal ellipse/parabola for a Gregorian system) is then the most obvious choice for base profiles.

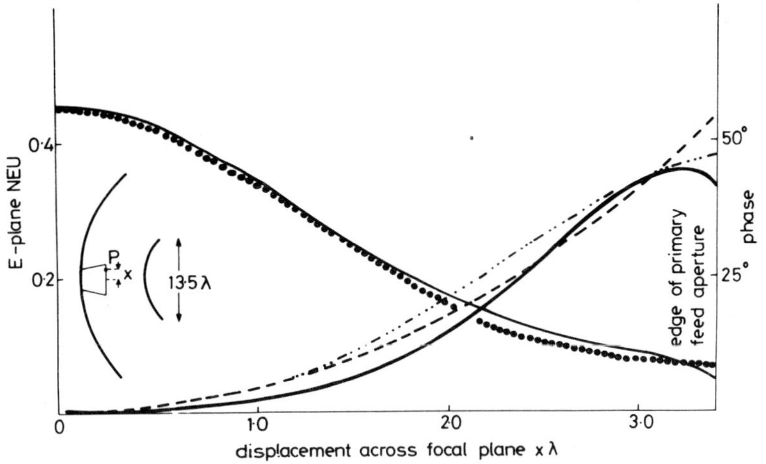

Fig. 7.15. *Receiving and transmitting fields in focal plane of DPS Cassegrain*

——— fully shaped } amplitude
• • • partially shaped }
—···— fully shaped } phase advance
——— partially shaped }
– – – – transmitting phase delay at primary feed aperture

However, when $D_s > 25\,\lambda$ the synthesis iterations will tend to converge rather slowly from a conic section starting point. Under these conditions, a worthwhile saving in computer time is to be had by substituting suitable modified Williams profile perturbations. It is often unnecessary to generate an explicit Williams solution for the actual geometry under study. Rather, it suffices to 'borrow' results from an existing Williams design of very roughly similar geometry. A normalised angle scale (e.g. $\psi_s/\hat{\psi}_s$) is used to translate the profile perturbations to the actual geometry (e.g. see Fig. 7.4).

GO ray tracing would suggest that a base hyperboloid should extend to an angle θ_C equal to the base paraboloid semi-opening angle θ_B (see Fig. 7.13). However, vector diffraction calculations demonstrate that when the antenna is receiving, the power focused by the main reflector will spread substantially beyond the GO ray cone limit $\theta = \theta_B$. According to one viewpoint, the shaped Cassegrainian system gains some of its advantage by capturing this extra power and transferring it in a correlated fashion to the feed horn.

To achieve this it is necessary to 'extend' the subreflector. The optimum extension ordinarily corresponds to the condition where the phase discrepancy between *E*- and *H*-plane receive patterns has reached about 90° at the rim of the subreflector, and θ_C is perhaps $1{\cdot}25\ \theta_B$.

7.5.2 Minimum blockage condition

The diameter of the effective overall main reflector central shadow is determined by whichever is the larger of the subdish shadow θ_A and feed shadow $\theta_{A'}$ of Fig. 7.13. To ensure that a specified proportion of the energy radiated by the feed is captured by the subreflector, the feed horn aperture size will need to be increased if the subreflector size is reduced. Thus, the overall GO blockage shadow will ordinarily be minimised when the feed and subreflector shadows are equivalent, e.g.

$$\theta_{A'} = \theta_A \tag{7.23}$$

The actual diameter value D_{min} corresponding to this condition can be established by trial and error calculations, evaluating receive and transmit fields at the subreflector surface in the manner already described. Suppose geometrical constraints are imposed as follows[7]:

(*a*) A main reflector opening semiangle of $\theta_B = 80°$
(*b*) A feed horn aperture positioned midway between the main reflector vertex and its base paraboloid focus
(*c*) A feed horn of corrugated conical type, with rear flange flush with the main reflector vertex.

These conditions result in the variation of D_{min} with final aperture diameter illustrated in Fig. 7.16.

Within the usual range of practical alternatives, a modification to either of the first two constraints listed above will not significantly affect the final D_{min} result. However, if the feed aperture is displaced as far as the surface of the main reflector, the antenna configuration tends to degenerate to a 'plane wave feed Cassegrain' form. Under these circumstances, the subreflector will be slightly larger than the feed horn for optimum η_t. The corresponding two blockage shadows are no longer equal, and D_{min} increases some 20% to 40% relative to Fig. 7.16.

In practice of course the effective shadowing diameter for a feed horn will be larger than the radiating part of its aperture. The curve of Fig. 7.16 therefore needs to be appropriately augmented to allow for

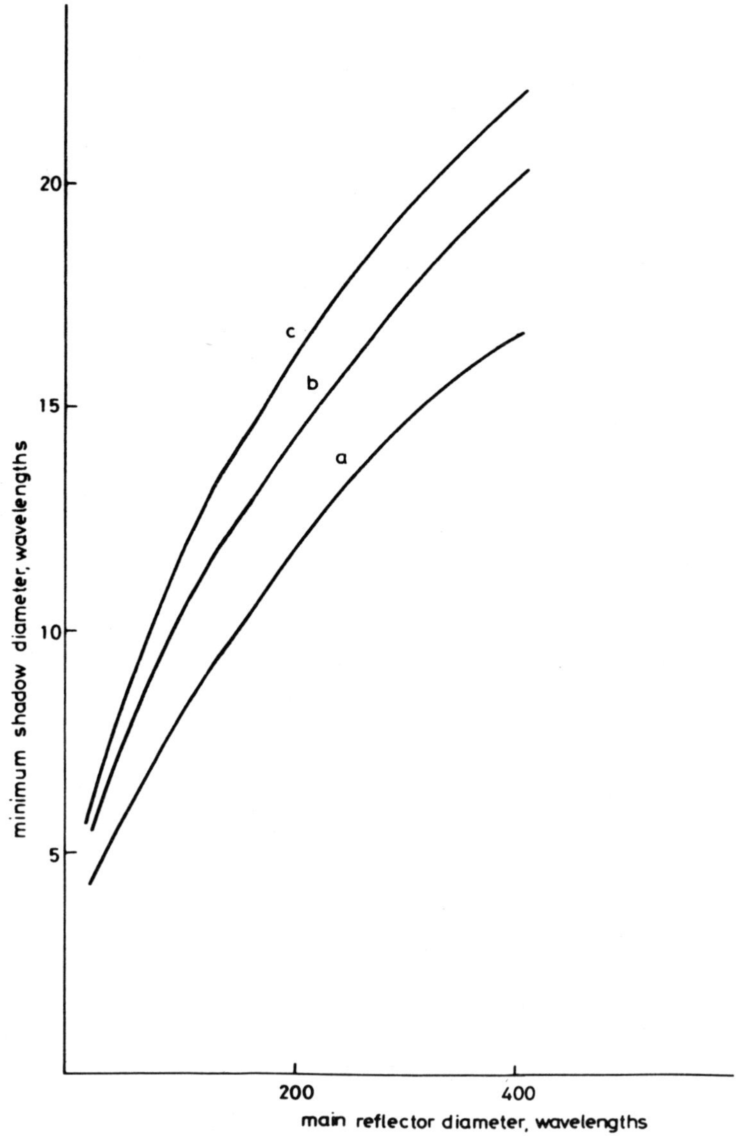

Fig. 7.16. *Minimum shadow diameter*
a hyperbola/parabola
b modified Williams
c DPS

horn wall thickness. For corrugated horns, the existence of grooves ensures an effective thickness in excess of $\lambda/4$.

7.5.3 Gain performance

The theoretical efficiency η_t of a Cassegrain antenna is fundamentally limited by the precision to which it is possible to obtain an ideal constant phase and amplitude, zero spill-over illumination of the final aperture. This in turn relates to the degree to which the compromise between beam-shaping and aperture blockage properties of the feed system can be resolved.

Curves for η_t calculated with simplifying assumptions, e.g. 'far field' approximations in both synthesis and analysis operations and a zero flare angle horn model, were presented in Reference 7. The curves of Fig. 7.17 reflect the results of more rigorous near field calculations,[8] utilising the feed horn model of Section 5.1.1. Strictly, they relate specifically to the geometrical constraints that were stipulated in Section 7.5.2. However, alternative DPS geometries having the same final aperture rarely differ significantly in respect of η_t. For plane wave feed designs the η_t values usually lie slightly below the curves, but the discrepancy can be explained directly in terms of increased aperture blockage.

By way of contrast, it is not possible to quote a universal, geometry-independent η_t curve for a Williams design. The only inviolable rule is that the efficiency will be less than that for DPS profiles. The performance of Williams profiles relates to the highly variable influence of finite range effects in a ray optics design, as discussed in Section 7.1.3.

The overall gain for the shaped Cassegrain will inevitably be subject to 'loss budget amendments' not included in the theoretical η_t given in Fig. 7.17 e.g.

(i) Subreflector profile and reflector alignment manufacturing tolerances
(ii) Heat losses in the feed horn, feeder waveguide, and microwave components
(iii) Aperture blockage and scattering from subreflector support struts
(iv) Gaps between reflector panels
(v) Losses due to non-ideal feed pattern, e.g. unbalance between *E* and *H*-plane patterns or excess of power in sidelobes in relation to the dual mode conical horn model used in deriving Fig. 7.17
(vi) Feed V.S.W.R. losses.

In connection with the first item listed, it is relevant to draw attention here to some of the wider implications of profile/alignment

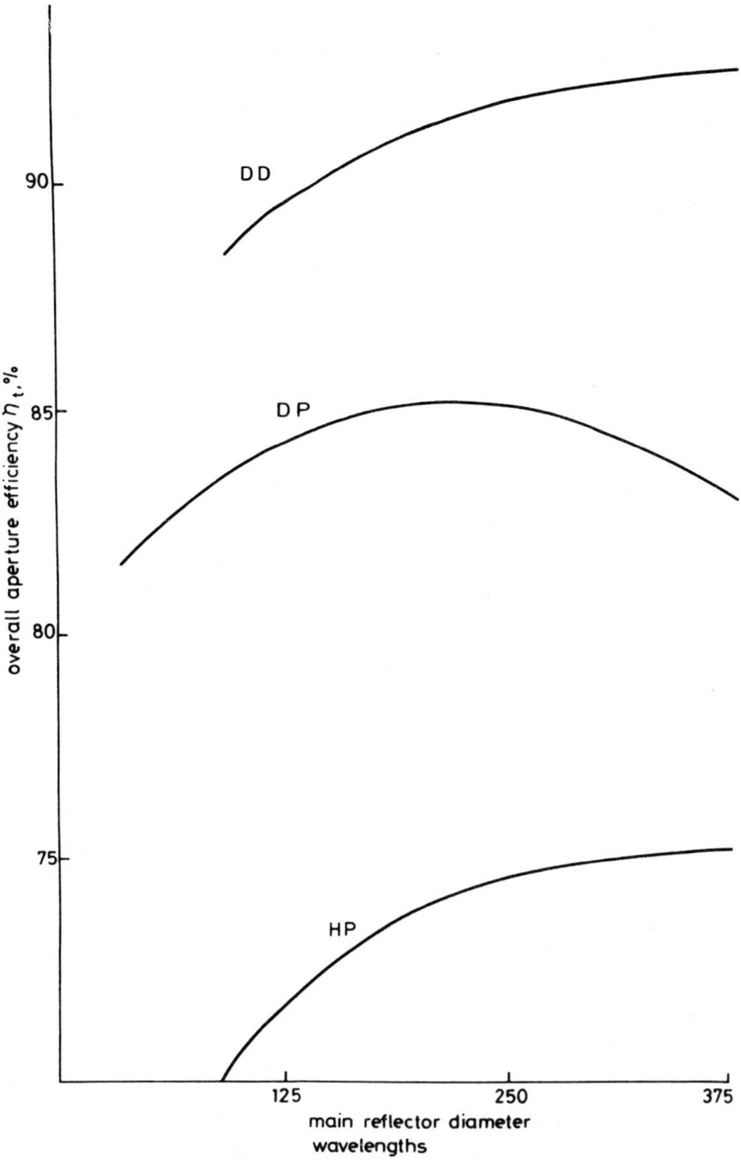

Fig. 7.17. *Comparison of efficiencies*
HP = hyperbola/parabola
DP = partially shaped DPS
DD = fully shaped DPS

errors. If mechanical limitations for a particular antenna force the designer to accept fairly severe errors, then some of the conclusions stated in the previous parts of this chapter may be modified. This is because the resulting illumination phase errors, *when combined with phase errors inherent in the nominal design*, could tend to magnify the deleterious effects of the latter (e.g. gain degradation is proportional to $(\text{phase error})^2$).

Specifically, on a 'worst case' basis, the performance deficit of Williams-derived nominal profiles in relation to DPS nominal profiles is augmented in the presence of severe profile errors. Similar reasoning applies in relation to a comparison of fully shaped profiles with the partially shaped versions to be discussed in the next Section.

7.6 Alternative shaped Cassegrain configurations

7.6.1 Partial shaping

In the partially shaped Cassegrain antenna, only the subreflector is perturbed relative to classic conic section profiles. One application is to update an existing hyperbola/parabola Cassegrain by retrofitting a new subreflector. Alternatively, it may be desirable to make use of an 'off-the-shelf' or existing main reflector.

With the DPS technique the necessary constraint is readily accommodated, simply by omitting the main reflector profile perturbation steps in Section 7.2.2. The subreflector profile is now different from that for a fully shaped design. Inevitably, there will be some penalty in electrical performance in relation to a fully shaped system. Fig. 7.12 compares scatter patterns for a partially shaped subreflector with those of the fully shaped design discussed previously in connection with the 150 λ Cassegrain. It will be observed that partial shaping has resulted in some over-illumination near the centre of the main reflector. Moreover, a significant illumination phase error is now evident.

The efficiency η_t of the partially shaped 150 λ Cassegrain is 0·85, as compared with 0·91 for the fully shaped case. In general, however, the amount of phase error present and hence the efficiency degradation with partial shaping increases with the size of the subreflector in relation to the wavelength. This trend is illustrated in Fig. 7.17.

In the context of focal plane patterns, the penalty incurred for constraining the main reflector profile appears in a somewhat different guise. In response to a plane wave incident axially on the main reflector, even the partially shaped 150 λ design synthesises an almost perfect conjugate match to the transmitting amplitude and phase distribution at the feed aperture (Fig. 7.15). However, the deficiency relative to full shaping now comes about via an increase in the level of sidelobes for the receiving focal plane pattern.

The partially shaped Cassegrain represents only one of many possible applications in which DPS profile synthesis is adapted to accommodate a design constraint. Another form of design constraint, in relation to subreflector matching, will be discussed in Chapter 9. However, the further implications are beyond the scope of this book.

Fig. 7.18. *19 m beam Cassegrain*

7.6.2 Antenna with more than two reflectors

In previous sections of this chapter the concepts of DPS synthesis have been expounded in terms of the standard two reflector Cassegrain system, this being the application which is both of the greatest practical importance, and also the most likely to be familiar to the reader. However, the use of the DPS technique is by no means restricted to this one particular case. For example, to treat a Gregorian geometry, it is merely necessary to substitute an ellipse for the more usual hyperbola as a starting point for the subreflector profile. Moreover, the DPS concept readily extends to the case of an antenna having more than two reflectors.

For a multiple reflector antenna, the *optimisation* operations of Section 7.2.2. remain unaltered. The only change is in the implementation of the *analysis* operations. Thus in general the evaluation of 'receive' or 'transmit' fields at the surface of a reflector whose profile is to be optimised will now involve a multiple scattering process. This in

Fig. 7.19. *Beam waveguide assembled for field probing*

turn can be broken down into individual near field scattering problems which can be solved successively by applying the S.W.E. techniques of Chapter 3.

7.7 19 m fully shaped beam Cassegrain antenna

Fig. 7.18 gives an exterior view of an 11/14 GHz satellite communications earth station based on a 19 m reflector[10, 11] and capable of working in 'frequency re-use mode' with orthogonal linear polarisations. A beam waveguide feed is used (Sections 5.1.5, 6.2.5).

The design of the Cassegrainian reflector system is based on the concept of an 'equivalent feed'. The latter might be envisualised as a hypothetical horn placed at the ray optics focus 0 of Fig. 5.10. However, the S.W.E. numerical technique functions most efficiently and is most accurate for a match to experimental data[12] when the co-ordinate origin for the expansion is near to the centre of the radiating source.

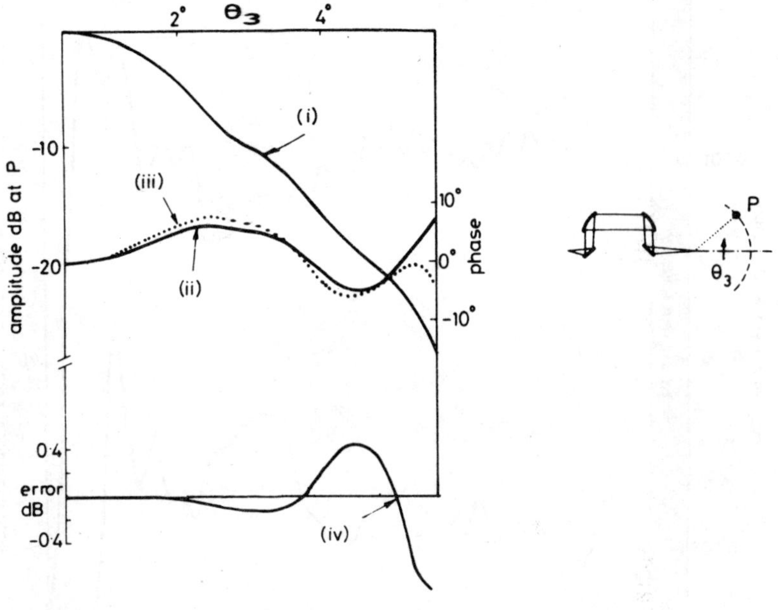

Fig. 7.20. *Feed system patterns for the beam waveguide (E-plane)*
(i) measured dB
(ii) measured phase
(iii) predicted phase
(iv) predicted-processed measurements

Thus the beam waveguide is replaced by an equivalent feed placed at the centre of its final reflector. This source is in turn represented as a Laurent series of spherical waves.

At the time at which the profile synthesis task was carried out, no experimental feed system data was yet available. The equivalent feed for DPS Cassegrainian synthesis was therefore based on a coaxial paraboloid analytical model of the kind described in Section 5.1.5. The centre used for the equivalent Laurent expansion is the vertex of the final coaxial paraboloid, corresponding to 0′ of Fig. 5.10.

A beam waveguide assembly tends to be too large and heavy to permit installation on a conventional antenna pattern range (Fig. 7.19). However, its electrical characteristics can be measured by a movable probe. In the scheme adopted, the probe moves on an arc of a spherical surface centred on a point near 0 of Fig. 5.10. This limits the phase deviation of the measured patterns to a relatively small value. A

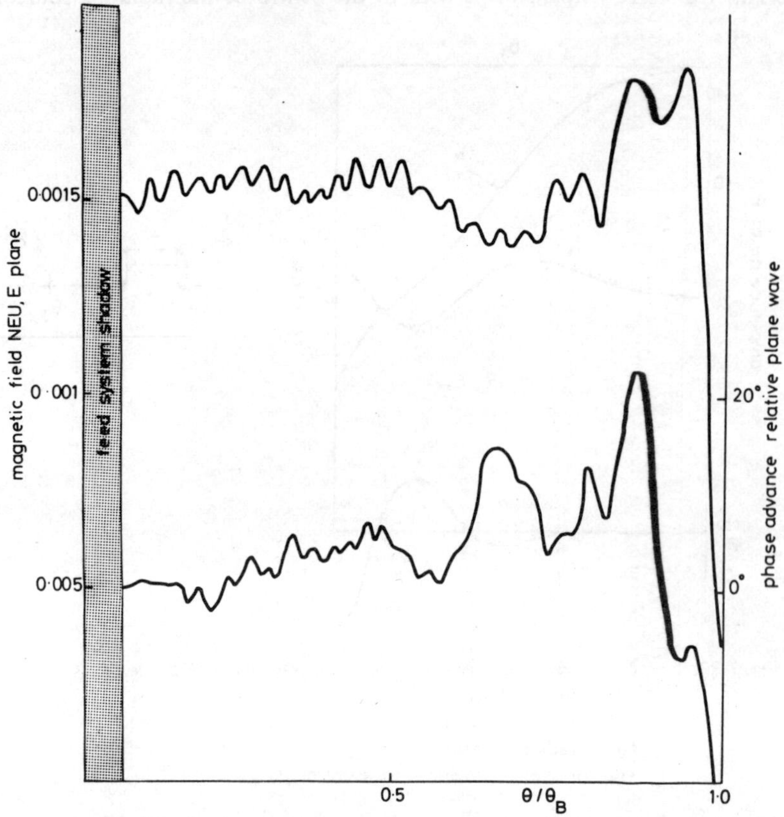

Fig. 7.21. *Predicted fields at main reflector surface as deduced from BWG field probing data*
upper curve: amplitude
lower curve: phase

Laurent S.W.E. centred on 0″, the midpoint of the final plane reflector of the beam waveguide, was matched to this data. The weighting coefficients were determined via eqn. 3.46 and the displaced sphere secondary wavefront formulation of Reference 12.

The near field phase and amplitude measurements confirm the prediction of the analytical model (e.g. see Reference 11 and Fig. 7.20). The empirically determined S.W.E. is used as a starting point for a final assessment of the electrical performance of the complete antenna. Thus

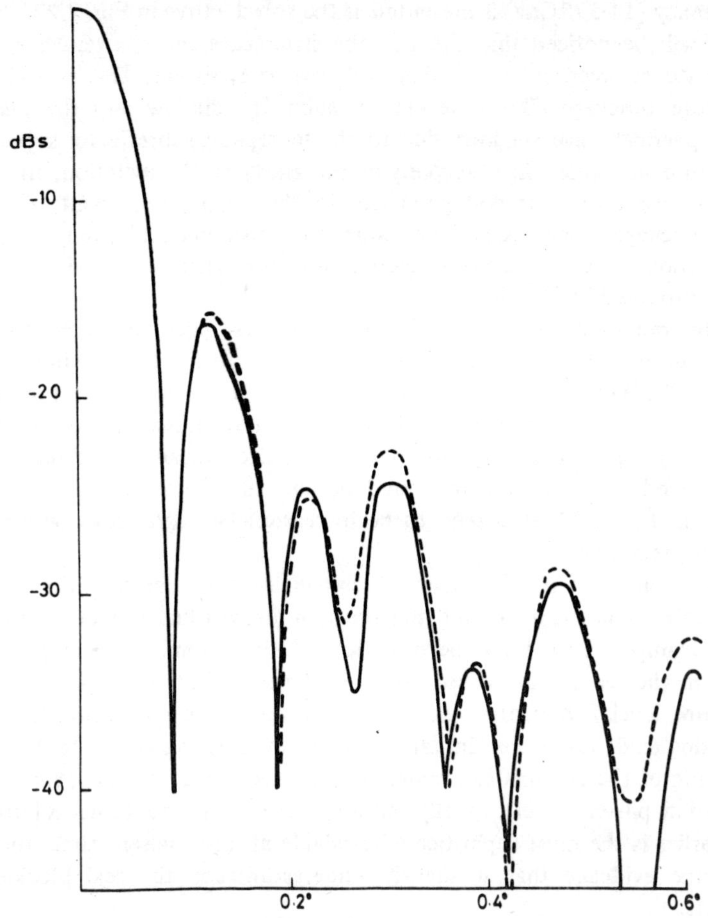

Fig. 7.22. *Predicted 11-575 GHz secondary pattern, effects of strut + subdish shadow included*
——— *theoretical*
– – – – *measured*

this S.W.E. is first expanded at the surface of the subreflector to facilitate the prediction of its PO induced currents. Another Laurent series then predicts the subreflector scattered pattern at the surface of the main reflector (Fig. 7.21).

The next step is to evaluate the PO currents that are induced on the main reflector surface. Finally, the secondary pattern of the complete antenna is calculated using the finite range KHS technique of Section 2.2.11. The predicted 34 km pattern at centre band receive frequency (11·575 Ghz) is presented as the solid curve in Fig. 7.22.

It will be noticed that the sidelobe structure shows a tendency to alternate between lobes of high and low amplitude. This is due to aperture blockage. Both the central subdish shadow and the plane and spherical wave shadows due to the tetrapod subreflector support structure are taken into account in the analytical prediction. In the present case, the tetrapod gives rise to the more important of the two blockage mechanisms. The model that has been adopted to treat diffraction by the rectangular section support struts is based on concepts proposed by Rusch.[13,14]

The radiation pattern of the antenna has been measured for a test site of 34 km propagation length (Goonhilly to Leswidden in Cornwall). Whereas the pattern ideally should have near perfect mirror image symmetry, mechanical tolerances for the Cassegrainian system inevitably cause small asymmetries. Thus, the mean of the left and right hand halves of the measured pattern is plotted as the broken curve in Fig. 7.22. It is seen to be in reasonable agreement with the earlier prediction.

One source of the few minor discrepancies that can be observed is probably small errors in the positions of individual reflector panels (for example, a ± 0·1 mm perturbation of the nominal 19 m profile can, in the worst case, cause up to 1 dB perturbation of the near-in sidelobe levels). A further factor entering into the comparison is the pointing angle resolution limitations of the measuring system (a 0·002° rotation of the antenna can represent a change of a decibel or more in radiation pattern level). Finally, although the tetrapod model referred to earlier is the most sophisticated available at the present time, there is some evidence that it slightly underestimates the real blockage effects.

References

1 GALINDO, V.: 'Design of dual reflector antennas with arbitrary phase and amplitude distributions', *IEEE Trans.*, 1964, AP-12, pp. 403–408

2 WILLIAMS, W. F.: 'High efficiency reflector antenna' *Microwave J.*, 1965, pp. 79–82

3 DAVEAU, B.: 'Synthesis and optimisation of antennas of special shape', *Rev. Tech. Thomson-CSF*, 1970, 2, pp. 37–55

4 ELLIOT, R. D., and POULTON, G. T.: 'Diffraction optimised shaped beam reflector antennas, *Electron. Lett.*, 1977, **13**, pp. 325–326

5 CLARRICOATS, P. J. B., and POULTON, G. T.: 'High efficiency microwave reflector antennas – a review', *Proc. IEEE*, 1977, **45**, pp. 1484–1502

6 WOOD, P. J.: 'Field correlation theorem with application to reflector aerial diffraction problems', *Electron. Lett.* 1970, **6**, pp. 326–327

7 WOOD, P. J.: 'Reflector profiles for the pencil beam Cassegrain antenna', *Marconi Rev.*, 1972, **35**, pp. 121–138

8 WOOD, P. J.: 'Near field defocussing in the pencil beam Cassegrain antenna', *Marconi Rev.*, 1973, **36**, pp. 201–223

9 WOOD, P. J.: 'Field correlation diffraction theory of the symmetrical Cassegrain antenna' *IEEE Trans.*, 1971, **AP-19**, pp. 191–197

10 BRAIN, J. R., et al.: 'An 11/14 GHz 19 m satellite earth station antenna', IEE International Conference on Antennas and Propagation, Nov. 1978. pp. 384–388

11 WOOD, P. J., et al.: 'A beam waveguide feed for an 11/14 GHz antenna, *ibid.*, pp. 223–229

12 WOOD, P. J.: 'The prediction of antenna characteristics from near field measurements, Part I Theory', *Marconi Rev.*, **40**, pp. 42–68

13 RUSCH, W. V. T.: 'Application of two dimensional integral equation theory to reflector analysis'. Jet Propulsion Laboratory report TM 33–478, May 15th, 1977

14 RUSCH, W. V. T.: 'Forward scattering from square cylinders in the resonance region with application to aperture blockage', *IEEE Trans.*, 1976, **AP-24**, pp. 182–189

Chapter 8

Focal plane patterns

8.1 Introduction

An investigation of field distributions in the focal plane often provides valuable insight into mechanisms of reflector operation. These distributions are the result of a near field scattering process, representing a condition where the antenna is operated in a receiving state, and responds to an incident plane wave. At first sight, one of the near field KHS vector diffraction integral forms of Section 2.1, in conjunction with the physical optics principle, would appear to offer a convenient and straightforward means of calculating the FPD. Minnett and Thomas[1] have adopted an approach of this type to evaluate the focal plane patterns of a rotationally symmetric parabolic reflector, whereas Bem[2] and Valentino[3] have analysed the offset parabolic reflector. However, in all these treatments, the simplicity of the basic diffraction integral (e.g. eqn. 2.2) is somewhat overshadowed by the series expansions and other special techniques used to reduce computer time requirements to manageable proportions. The resultant approximation tends to limit the generality of the solutions obtained: long focal length, observation points close to the optical focus point, a nearly on-axis incident wave etc.

GTD techniques do not appear to be applicable because of the focusing action of the reflector.[4] However, the S.W.E. method of Chapter 3 is ideally suited to focal plane calculations. A numerically efficient solution is obtained while introducing no source of approximation other than the physical optics principle.

For a front-fed reflector, the starting point for an FPD calculation is the PO induced currents for incident plane-wave excitation. It is usually most expedient to introduce an expansion of the Taylor type, centred on the reflector focus. An example of this type of implementation has been given in Section 3.3.4.

For a Cassegrain antenna, two successive scattering processes are involved in the formation of the focal plane distribution. The first of these can be treated via a Taylor series as described above.[5] In the Cassegrainian case, the Taylor series is used to determine the receiving fields incident on the subreflector, and hence ultimately the subreflector PO induced currents. The second stage of the calculation can employ either a Taylor or a Laurent series. For geometries having a secondary focus very close to the subreflector, it will be necessary to displace the origin of a Laurent series to the concave side to ensure that the focal plane lies entirely outside the minimum sphere. In any event, the Laurent series will be numerically less efficient than the Taylor series if the portion of the focal plane where it is desired to evaluate the fields is of smaller diameter than the subreflector.

In general, for either the front-fed or the Cassegrainian case, if the reflector is asymmetric or the received field is incident off-axis, a full S.W.E. mode set will be generated as in Section 3.3.4. Nevertheless, simple 'rotational symmetry' calculations can provide a surprisingly comprehensive insight into antenna performance characteristics. Information regarding intrinsically off-axis features such as beam deviation factor and crosspolar patterns can be obtained by S.W.E. calculations using only $m = 1$ modes.

8.2 Prime focus feed design

For a prime-focus fed reflector, the use of the focal plane technique to stipulate desirable feed horn properties and predict maximum achievable values of the antenna efficiency η_t is well established.[6, 7]

Robieux's theorem (see Section 4.1) applied to focal plane calculations specifies a transmission coefficient T,

$$T = \tfrac{1}{2} \int_{FP} \{\boldsymbol{E}_f \times \boldsymbol{H}_{FP} + \boldsymbol{H}_f \times \boldsymbol{E}_{FP}\} \cdot d\boldsymbol{S} \tag{8.1}$$

where $\boldsymbol{E}_f : \boldsymbol{H}_f$ is the feed transmitting field and $\boldsymbol{E}_{FP} : \boldsymbol{H}_{FP}$ the field scattered by the reflector for an incident plane wave. Both fields are normalised to 1 watt power.

It may readily be demonstated that $T = 1$ if

$$\boldsymbol{E}_f = \overset{*}{\boldsymbol{E}}_{FP} \tag{8.2}$$

and

$$\boldsymbol{H}_f = -\overset{*}{\boldsymbol{H}}_{FP} \tag{8.3}$$

These results are sometimes taken to imply a fundamental theoretical limit of $\eta_t = 1$ for a reflector antenna, corresponding to a case where a hypothetical feed system presents a conjugate match to receive fields over the entire focal plane. Equally, and more realistically, if the feed is to be constrained to have 'aperture limited' properties, then it is assumed that an optimum η_t is realised when a conjugate match holds over the specified restricted aperture.

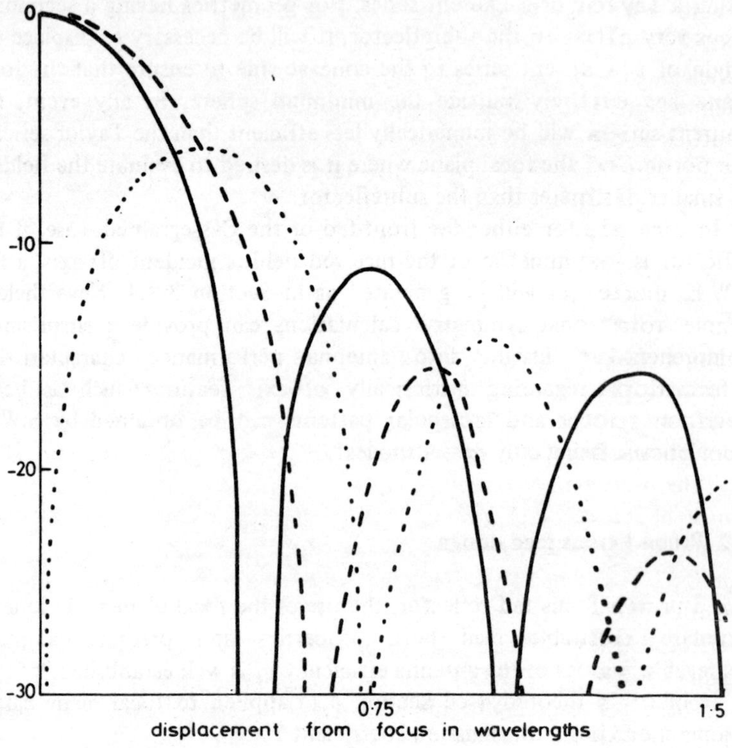

Fig. 8.1. *E-plane focal plane pattern F/D = 0·33 paraboloid*
- - - - transverse electric field
——— transverse magnetic field
· · · · · axial electric field

It does not appear possible to rigorously substantiate the first statement, and the second certainly does not hold precisely. Nevertheless, both are always valid to engineering accuracy in ordinary applications, with a possible exception relating to the crosspolar pattern in the restricted aperture case.

Fig. 8.1 gives the computed focal plane pattern for a 100 λ $F/D =$ 0·33 rotationally symmetric parabolic reflector. It will be observed that the E- and H-field patterns differ markedly away from the FPD maximum at the focal point. Specifically, $E/H > 1$ for an E-plane cut across the principal lobe, and $E/H < 1$ for an H-plane cut. This is just one aspect of the 'balanced hybrid' nature of these fields, a property first expounded by Minnett and Thomas.[1] The rigorous Taylor S.W.E. vector diffraction solution indicates that the FPD fields do not conform quite exactly[8] to the ideal balanced hybrid relationships of Section 3.2.7.2, and there is a second-order dependence on frequency. Nevertheless, the conclusion of Minnett and Thomas' approximate analysis,[1] namely an FPD dependent only on reflector opening angle, is valid to a sufficient degree of accuracy for most practical purposes.

A paraboloid FPD may also be resolved into its 'Huygens source components',

$$P_{co}^{FP} = \tfrac{1}{2}(\boldsymbol{E}_{FP} \cdot \boldsymbol{a}_x + \boldsymbol{H}_{FP} \cdot \boldsymbol{a}_y) \qquad (8.4)$$

$$P_{\times}^{FP} = \tfrac{1}{2}(\boldsymbol{E}_{FP} \cdot \boldsymbol{a}_y - \boldsymbol{H}_{FP} \cdot \boldsymbol{a}_x) \qquad (8.5)$$

Here x and y are measured in the focal plane with z drawn away from the reflector. In the particular case of ideal balanced hybrid conditions, eqn. 4 specifies a circularly symmetric copolar distribution. The principal lobe of P_{co}^{FP} extends to $r = 0{\cdot}52\,\lambda$ for the $100\,\lambda$ paraboloid.

The non-unity wave impedance associated with the FPD of Fig. 8.1 must be taken into account when calculating power flow. Then it is found that 68% of the available power is intercepted by a central disc of $0{\cdot}52\,\lambda$ radius. The remaining power resides in sidelobes. Predominantly, the direction of propagation in these sidelobes is away from the reflector, as is the case for the FPD principal lobe. However, at certain radii ($r = 0{\cdot}6\,\lambda, 1{\cdot}6\,\lambda, 3{\cdot}2\,\lambda$ etc.), a reverse power-flow anomaly occurs.

These considerations impose a limit on the antenna efficiency η_t that can be realised using a prime focus feed. One class of feed has aperture dimensions of the order of a wavelength and attempts to collect power only from the FPD principal lobe. Probably the most efficient device of this general type is the 'ICR' horn described in Chapter 6. For an ICR-fed paraboloid of optimum design, $\eta_t = 0{\cdot}66$.

A higher antenna efficiency can be achieved (albeit at the expense of increasing the complexity and aperture blockage of the feed structure) by using a wide angle corrugated conical horn[9] or 'scalar feed'. This type of horn extracts some of the power in the focal plane sidelobes. However, the smoothly varying amplitude and phase distribution corresponding to dominant mode excitation of its aperture cannot match alternative sidelobe peaks and nulls. Moreover, no equivalent of the reversed power flow regions of the FPD is provided.

In consequence, although η_t is somewhat enhanced relative to a principal lobe extraction feed, it will rarely exceed 0·75. A yet lower net figure will obtain if the diameter of the reflector is such that feed aperture blockage is significant.

A more efficient utilisation of the power in the FPD sidelobes is ideally possible with a multimode feed horn.[10, 11] However, multimoding usually leads to a very complex feed structure, increased blockage and narrow band operation. A more straightforward solution for large reflectors is to employ a Cassegrainian feed. The shaped subreflector reforms the FPD to reduce the power content of the sidelobes and eliminate the 'reversed propagation' regions (see Section 8.7 and Chaper 7).

8.3 Crosspolarisation in plane of prime focus

The balanced hybrid property implies that the electric-field vector distribution in the *H*-plane is similar to that of the magnetic field in the *E*-plane shown in Fig. 8.1. Thus the curves of Fig. 8.1 are representative of the discrepancy between *E*- and *H*-plane cuts for (say) the ***E***-vector. This discrepancy corresponds to a fairly substantial crosspolar level according to eqn. 6.7, amounting to −19 dB relative to copolar peak at 0·6 λ off-axis in the 45° plane. The *E*- and *H*-vector crosspolar signals are of almost equal amplitude, but are in an antiphase relationship with the sign conventions of Fig. 8.2.

If the prime focus feed has dipole-like polarisation characteristics, it will couple selectively to the ***E*** vector of the FPD, and hence will sample the electric FPD crosspolar component. The secondary pattern crosspolarisation then increases with increasing disparity of the ***E*** and ***H*** focal plane patterns, e.g. reducing *F*/*D*. Hence long focal length is beneficial with a dipole-like prime focus source.

More commonly, prime focus feeds approximate Huygens source (or Ludwig's third definition[12]) polarisation characteristics. There is then equal coupling to the ***E*** and ***H*** field components of the FPD, resulting in an effective near cancellation of the corresponding crosspolar signals. In fact, there remains only a small reflector depolarisation lobe in the secondary pattern, as described in Section 6.2.4. This behaviour may also be inferred from eqn. 5, in that the 'Huygens source component' FPD crosspolar pattern will be suppressed to the extent that the paraboloid launches balanced hybrid fields in the focal plane.

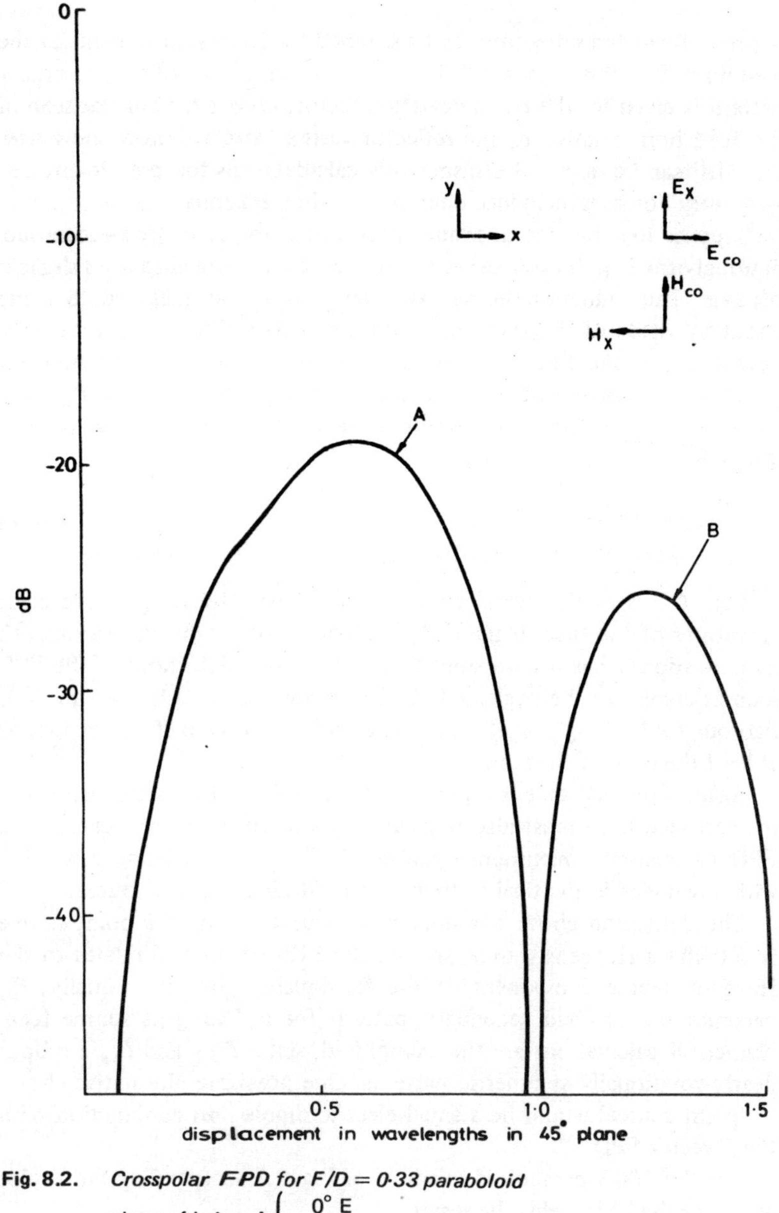

Fig. 8.2. *Crosspolar FPD for F/D = 0·33 paraboloid*

phase of lobe A $\begin{cases} 0° \text{ E} \\ 180° \text{ H} \end{cases}$

phase of lobe B $\begin{cases} 0° \text{ H} \\ 180° \text{ E} \end{cases}$

8.4 Beam deviation factor

A prime focus-fed reflector may be scanned by lateral movement of the feed horn. For limitingly small displacements, the scan of the secondary pattern is given by the beam deviation factor, times the angular scan of the feed horn relative to the reflector vertex. We will now show how the BDF can be deduced utilising *only* calculations for 'pseudo-circular symmetry', on-axis incidence operation of the antenna.

Suppose that the actual prime focus feed is replaced by an idealised, limitingly small 'point source'. A far-field probe is now displaced slightly off-axis. The reduction in received signal may be imagined to come about by virtue of the associated scan Δx of the FPD, the point source now sampling the FPD off-axis. By way of contrast, in the transmit mode the far-field probe is misaligned with the peak of P_{ps}, the secondary pattern with point source feed. Thus applying reciprocity principles,

$$\frac{P_{FP}(\Delta x)}{P_{FP}(0)} = \frac{P_{ps}(\text{BDF}\ \Delta x/F)}{P_{ps}(0)} \tag{8.6}$$

Eqn. 6 relates the beam deviation factor to the ratio of boresight curvatures of the focal plane and point source patterns. In carrying out its derivation it has been assumed that any scan distortions of the FPD such as coma may be neglected. This assumption (and hence eqn. 6) is rigorous for limitingly small scan, since such distortion effects reduce as at least the cube of the scan angle.

Again, provided that it is permissible to neglect FPD scan distortion, the real feed horn must also translate in sympathy with the scan of the FPD to maintain maximum signal level. Thus the limiting case BDF with a real feed is identical to that with an idealised point source.

The derivation above has implicitly assumed that the point source is actually a Huygens source, so that the LHS of eqn. 6 relates to the Huygens source component of the focal-plane pattern. Equally, P_{ps} becomes the far-field secondary pattern for a Huygens source feed. Numerical calculations are then simplified, since P_{FP} and P_{ps} are both nearly rotationally symmetric patterns. One possible alternative choice of 'point source' would be a small electric dipole, in conjunction with the E-vector FPD.

For the 100 λ paraboloid example of Fig. 8.1, expanding the Taylor series for the FPD yields the result

$$\frac{P_{FP}(\Delta x)}{P_{FP}(0)} = 1 - 8{\cdot}72 \times 10^{-3}$$

for $\Delta x = 0{\cdot}04\,\lambda$. Furthermore, the Laurent series for the Huygens source secondary pattern specifies

$$\frac{P_{ps}(\Delta x/F)}{P_{ps}(0)} = 1 - 1{\cdot}547 \times 10^{-2}$$

Taken together, these two calculations imply a BDF of 0·751. Comparable published results are 0·76 with 10 dB illumination taper and 0·742 with uniform illumination (Ruze[13]), or 0·775 with 10 dB illumination taper (Ingerson[14]). The small discrepancies and apparent dependence on illumination taper here appear to be associated with the 'finite displacement' evaluations used by the two referenced authors.

8.5 Focal plane patterns for a hyperbola/parabola Cassegrain antenna

Fig. 8.3 gives the computed secondary focus FPD for a small hyperbola/parabola Cassegrain antenna with a 33 λ main reflector of 90° opening semiangle, and a 5 λ subreflector. This pattern has somewhat lower sidelobes (e.g. −16·3 dB first) than the prime focus case of Fig. 8.1. Nevertheless, at least for the particular geometry selected, the power in the principal lobe is only 68% of the power intercepted by the main reflector.

This latter result takes into account shadowing of the secondary aperture by the subreflector, which accounts for part of the power loss. A loss can also be attributed to edge diffraction at the subreflector. Thus for much larger hyperbola/parabola geometries, the relative power content of the FPD principal lobe can approach 80%. In any event, there is no evidence of reversed propagation sidelobes for Cassegrainian reflector systems, and usually only marginal semblance to the ideal balanced hybrid interrelationship of *E*- and *H*-FPD fields discussed in Section 8.2.

The 45° plane crosspolar FPD for the small Cassegrain is given in Fig. 8.4. In contrast to the prime focus case, it will be observed that the electric and magnetic vector patterns not only are of different amplitude levels but are essentially cophasal rather than antiphased. It may be deduced that the *intrinsic* crosspolarisation suppression property of a prime radiator consisting essentially of an ensemble of Huygens sources with 377 Ω aperture impedance applies *only* to its use as a front feed (see also the end of the next Section).

These depolarisation phenomena are a consequence of vector diffraction at the electric wall, finite diameter, subreflector. Ideally, corrugations could be added to the surface of the 5 λ subreflector to attempt to reduce the disparity between electric and magnetic vector scattering properties. However, the surface illumination and wave incidence angles are quite different to that obtaining in a corrugated conical horn, and corrugations on a subreflector do not appear to be effective in practice.[15]

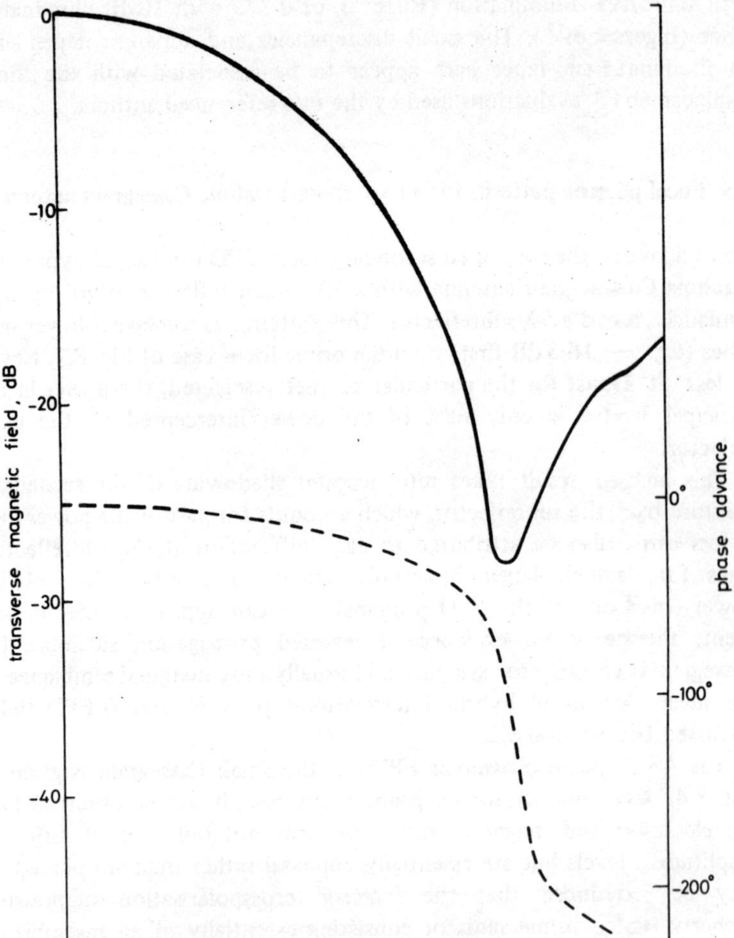

Fig. 8.3. *E-plane focal plane pattern, 5 λ hyperboloid*
——— amplitude
- - - - phase

8.6 Calculation of Cassegrain secondary pattern crosspolarisation

The preceding discussions have tacitly assumed the existence of an explicit relationship between FPD and secondary pattern cross-polarisation. Thus, following the approach of Section 8.4, suppose that

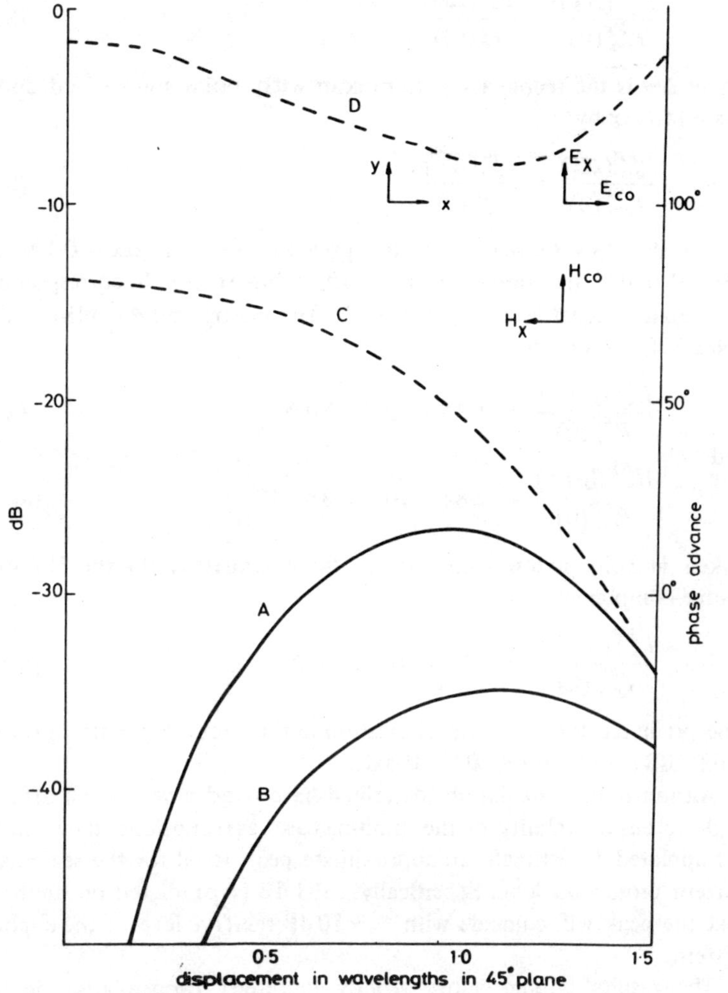

Fig. 8.4. *Focal plane crosspolar pattern, 5 λ hyperboloid*
A amplitude E-field
B amplitude H-field
C phase E-field
D phase H-field

the real feed horn is replaced by a small 'point source' sampling the centre of the focal plane pattern. The far field probe is then aligned so as to sense antenna crosspolarisation. On displacing the point source off-centre by an amount Δx in the 45° plane, the relative antenna response is

$$\frac{P_{x}^{FP}(\Delta x)}{P_{co}^{FP}(0)} = \frac{P_{x}^{ps}(\Delta\Theta)}{P_{co}^{ps}(0)} \tag{8.7}$$

where $\Delta\Theta$ is the secondary pattern scan with point source feed, and is related to Δx by

$$\frac{P_{co}^{FP}(\Delta x)}{P_{co}^{FP}(0)} = \frac{P_{co}^{ps}(\Delta\Theta)}{P_{co}^{ps}(0)} \tag{8.8}$$

For the 33 λ Cassegrain of the previous section, $\Delta x = 0{\cdot}1\,\lambda$ and $\Delta\Theta = 0{\cdot}0204°$ are compatible scans according to eqn. 8, corresponding to a relative signal level of $-0{\cdot}119$ dB. The computed 45° plane cross-polar FPD vectors are

$$\frac{E_{x}^{FP}(0{\cdot}1\,\lambda)}{P_{co}^{FP}(0)} = 1{\cdot}31 \times 10^{-3} \angle 80{\cdot}6° \tag{8.9}$$

and

$$\frac{H_{x}^{FP}(0{\cdot}1\,\lambda)}{P_{co}^{FP}(0)} = 2{\cdot}84 \times 10^{-4} \angle 141{\cdot}4° \tag{8.10}$$

Taken in conjunction with eqn. 5, these results yield the Huygens source component

$$\frac{P_{x}^{FP}(0{\cdot}1\,\lambda)}{P_{co}^{FP}(0)} = 7{\cdot}33 \times 10^{-4} \angle 90{\cdot}3° \tag{8.11}$$

The predicted level of crosspolarisation in the secondary pattern is thus $-62{\cdot}7$ dB $\angle 90{\cdot}3°$ at $0{\cdot}0204°$ off-axis.

Although the calculations described have used a very small off-axis angle to ensure validity of the 'limiting-case' assumptions, they can be extrapolated to estimate an approximate peak level for the secondary pattern crosspolar lobe. Specifically, -31 dB is predicted on the basis that the peak will coincide with $\triangleq -10$ dB relative level of the copolar pattern.

The results quoted above strictly represent crosspolarisation (or depolarisation since the feed has zero Ludwig 3 crosspolar pattern) for a feed element giving almost isotropic illumination of the sub-reflector. For the more practical case of $\simeq -10$ dB taper, the level of the depolarisation lobe (as may be calculated by methods described in Chapter 6) is actually reduced by about 2 dB.

The focal plane approach can be used to investigate the suitability of different types of feed horn in respect of Cassegrain crosspolar performance. For example, eqns. 9 and 10 suggest that a magnetic dipole-like (e.g. high impedance waveguide aperture) characteristic would ideally be better than a Huygens source characteristic. In fact, this conclusion holds equally for quite large antennas; however, for a large antenna other considerations are more important in practice.

Some numerical data for a large Cassegrain design will serve to clarify the latter point. The feed horn is of 11 λ aperture diameter and 3·4° flare semiangle. Suppose that in some artificial manner its aperture were made to radiate as a planar array of *parallel* magnetic dipoles. Then according to the superposition principle of Section 6.2.3, its crosspolar pattern is a (narrow pencil beam) 'array factor' $f(\theta)$ multiplied by the (wide beam) crosspolar pattern for a magnetic dipole 'element'. From eqn. 6.9, in the 45° plane the resulting product is

$$\tfrac{1}{2}(1-\cos\theta)f(\theta)$$

This may be expected to maximise at about the −10 dB point of $f(\theta)$. Allowing for the finite range of the subreflector, an off-axis angle of 6·2° and a maximum relative level of −60·5 dB for the product is implied. Furthermore, after scattering from the Cassegrain reflectors, the relative level of the secondary pattern signal due to magnetic dipole sources, available to cancel depolarisation owing to the finite diameter of the subreflector, will be $\simeq -65$ dB.

It may be concluded that for a feed aperture as large as 11 λ, any question as to whether the aperture radiates as Huygens sources or dipoles is of little consequence. The prime consideration in practice must be the extent to which all dipoles or Huygens sources are aligned truly *parallel* to the nominal direction. It is for this reason that the corrugated conical feed horn, although essentially a Huygens source radiator (see Reference 4 and 5 of Chapter 5) is unlikely ever to be superseded for low crosspolarisation, narrow beam applications requiring substantial bandwidth.

8.7 Focal plane patterns for the shaped Cassegrain antenna

When high efficiency 'shaped' profiles are used in a Cassegrainian system, the focal plane sidelobes are found to be substantially suppressed.[8] Moreover, the phase error relative to transmitting illumination of the feed aperture is almost eliminated. In a particular design, a 41 λ subreflector forms the FPD of Fig. 8.5, the principal lobe being specified by $r < 6{\cdot}83\,\lambda$, with a power content of 0·925 watts. This figure allows for aperture blockage. For 1 watt incident on the 550λ main reflector, the reflected power captured by the subreflector in receive mode is 0·969 watts.

The Cassegrain profile optimisation was carried out with a feed horn model corresponding to a TE_{11}/TM_{11} conical horn of $5{\cdot}5\,\lambda$ internal radius. Its aperture encompasses 0·907 watts of FPD power. The horn transmitting aperture illumination presents an excellent conjugate match to the receiving FPD, permitting 0·89 watts to be extracted. The predicted overall antenna efficiency is hence 89%.

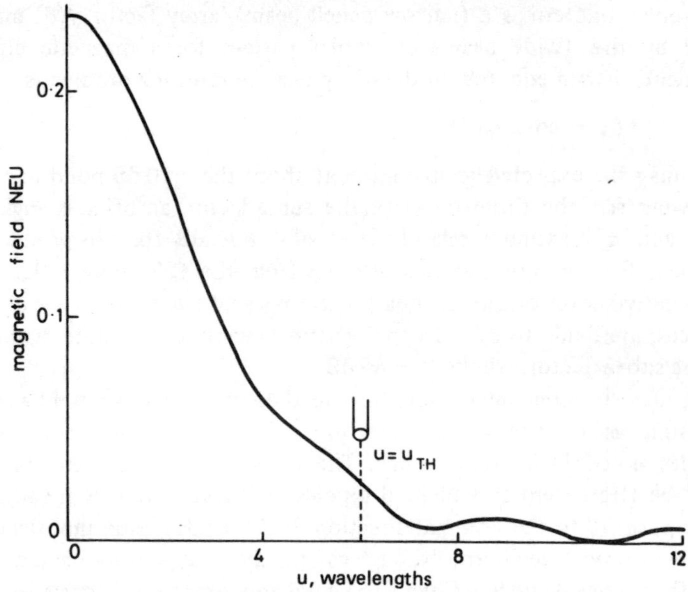

Fig. 8.5. *Focal plane pattern for a shaped Cassegrain antenna*

8.8 Tracking feed calculations

Many properties of dual reflector monopulse or autotrack feed systems can be related to a Cassegrainian beam deviation factor. The latter may be defined as the ratio of secondary pattern scan to feed horn aperture scan, the latter being measured at the focus of the main reflector.

For the Cassegrain of Section 8.7, the feed horn aperture lies in a 'focal plane' displaced by $F' = 122{\cdot}2\,\lambda$ from the main reflector focus.

From eqn. 6 and the computed patterns of Figs. 8.5 and 8.6, the beam deviation factor is specified as

$$(\mathrm{BDF})_D = 0{\cdot}0577 \tag{8.12}$$

A beam deviation factor could also have been estimated very approximately by invoking the equivalent paraboloid concept. The subreflector

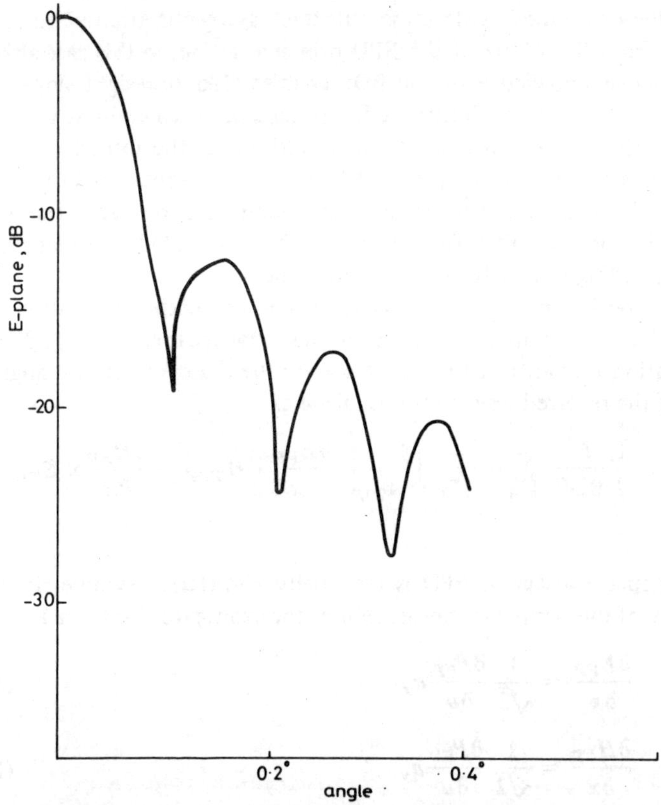

Fig. 8.6. *Point source secondary pattern for a shaped Cassegrain antenna*

subtends $\pm 9{\cdot}92°$ at the centre of the feed horn aperture. Thus the equivalent 550 λ prime focus-fed reflector with a corresponding opening angle has a 1584 λ focal length. The beam deviation factor for this equivalent paraboloid would be virtually unity. On this basis the geometrical optics estimate for the Cassegrain BDF would be

$$(\mathrm{BDF})_{GO} = \frac{122{\cdot}2}{1584} = 0{\cdot}0771 \tag{8.13}$$

The discrepancy between eqns. 12 and 13 is not unexpected as only eqn. 12 takes into account vector diffraction and the effect of shaped reflector profiles. As with the prime focus case of Section 8.4, it is significant that to evaluate the limiting case scan performance while incorporating these latter features, only calculations of the 'pseudo-circular symmetry' type of Section 2.2.11 are required.

Some tracking arrangements such as four horn monopulse systems or higher-order mode extraction autotrack systems attempt to capture power from the whole of the FPD principal lobe. In this case although in principle knowledge of the BDF enables the boresight slope of the tracking pattern to be determined, it is necessary to evaluate an integral numerically to obtain a solution. Specifically, the integral is that of Robieux's theorem (see Section 4.1), in the present case in respect of the feed horn transmitting aperture illumination, and the spatial derivative of the FPD. The latter is to be evaluated in a plane of cut corresponding to that for the tracking slope.

The focal plane approach leads to a more explicit solution when applied to the peripheral horn autotrack arrangement of Fig. 8.7. The correlation methods of Chapter 4 specify the output per unit angle off-axis of the received wave in the xz-plane as

$$\frac{1}{2}\frac{F'}{\text{BDF}}\sum_{1,4}-\sum_{2,3}\iint_{A_{TH}}\left\{\frac{\partial \boldsymbol{E}_{FP}}{\partial x}\times \boldsymbol{H}_{TH}-\frac{\partial \boldsymbol{H}_{FP}}{\partial x}\times \boldsymbol{E}_{TH}\right\}\cdot d\boldsymbol{S} \tag{8.14}$$

The shaped Cassegrain FPD is essentially circularly symmetric in the vicinity of the peripheral horns. Hence, for example, for horn 1

$$\frac{\partial \boldsymbol{E}_{FP}}{\partial x}=\frac{1}{\sqrt{2}}\frac{\partial P_{FP}}{\partial u}\boldsymbol{a}_x \tag{8.15}$$

and

$$\frac{\partial \boldsymbol{H}_{FP}}{\partial x}=\frac{1}{\sqrt{2}}\frac{\partial P_{FP}}{\partial u}\boldsymbol{a}_y \tag{8.16}$$

These latter results give a transformation to FPD slopes measured in the plane of a cut through beam centre. Usually, the peripheral horns will be of sufficiently small aperture area A_{TH} that the slopes involved may be taken to be constant quantities. The tracking efficiency K (defined as tracking volts per degree per 'communication' volt) can then be deduced as[16]:

$$K=\left(\frac{2\eta_{TH}A_{TH}}{\eta_{com}}\right)^{1/2}\frac{F'}{\text{BDF}}\left|\frac{\partial P_{FP}}{\partial u}\right|_{u=u_{TH}} \tag{8.17}$$

Here η_{TH} is the aperture efficiency of a single tracking horn operated as a transmitting antenna in its own right, and η_{com} is the overall boresight antenna efficiency in respect of the communications signal.

A particular peripheral horn tracking feed for the Cassegrain antenna of Section 8.7 uses four dominant mode open-ended circular waveguides as radiators, and has parameters

$$A_{TH} = 0{\cdot}5\,\lambda^2$$

$$\eta_{TH} = 0{\cdot}9$$

and

$$\left|\frac{\partial P_{FP}}{\partial u}\right|_{u=u_{TH}} = 0{\cdot}0187\sqrt{\text{watts}}/\lambda^2,$$

for 1 watt on main reflector

Eqn. 17 then specifies a tracking efficiency K of 0·69 volts/volt/°. That is, an antenna misalignment of say 0·001° will be sensed by the auto-track system as a signal of 0·001 x 0·69 volts assuming that it is amplified by an amount which would be sufficient to yield a 1 volt signal in the communications channel.

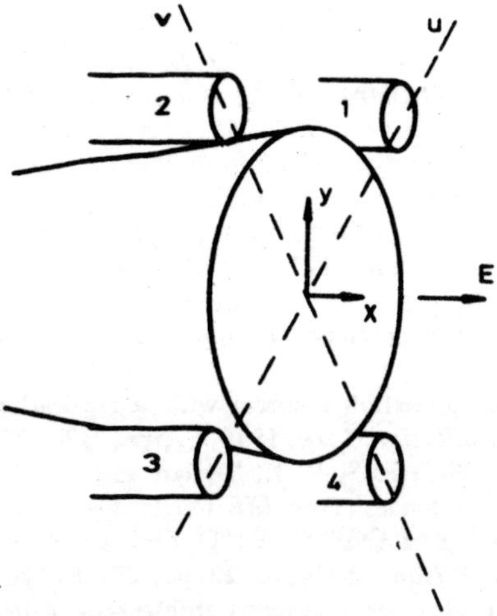

Fig. 8.7. *Peripheral horn tracking feed*

The advantage of calculating K in this way is that two-dimensional integrations are avoided. Thus in transmitting mode the peripheral horn assembly launches a kind of 'wide base line interferometer pattern'. Attempting to treat the scattering of this pattern by the Cassegrain subreflector and main reflector would inevitably result in an excessive amount of numerical working.

The focal plane approach lends itself to the instigation of a methodical design procedure for peripheral tracking feeds. Thus for example:

(*a*) Increasing the aperture size of the tracking horns will increase K only until 'negative slope' regions of an FPD such as Fig. 8.5 are encountered. Moreover, positioning the horns too close to a negative slope region can result in a multilobe tracking system secondary pattern with small lobes on either side of boresight and larger lobes further off-axis. This anomaly[17] is usually more accentuated with hyperbola/parabola as opposed to shaped profiles, because of the relatively high level of the FPD first sidelobe in the former case.
(*b*) Placing the horns at reduced radius will result in intercepting a region of the FPD of increased slope, but necessitate a smaller, less efficient, central communications feed.
(*c*) Focal plane calculations can be used to establish the relative merits of alternative feed cluster geometries in a given application.

References

1 MINNETT, H. C., and THOMAS, B. M.: 'Fields in the image space of symmetrical focusing reflectors', *Proc. IEE* 1966, **115**, (10), pp. 1419–1430

2. BEM, D. J.: 'Electric field distribution in the focal region of an offset paraboloid', *ibid.*, 1969, **116**, (5), pp. 679–684

3 VALENTINO, A. R., and TOULIOS, P. P.: 'Fields in the focal region of offset parabolic antennas', *IEEE Trans.*, 1976, **AP-24**, pp. 859–865

4 CLAYDON, B., and JAMES, G. L.: 'On the evaluation of image region fields for dual reflector scanning antennas'. European Microwave Conference 1974, pp. 343–347

5 WOOD, P. J.: 'Field correlation theorem with application to reflector aerial diffraction problems', *Electron. Lett.*, 1970, **6**, pp. 326–327

6 RUDGE, A. W., and WITHERS, M. J.: 'Design of flared horn primary feeds for parabolic reflector antennas', *Proc. IEE* 1970, **117**, (9), pp. 1741–1749

7 TRUMAN, W. M., and BALANIS, C. A.: 'Optimum design of horn feeds for reflector antennas', *IEE Trans.*, 1974, **AP-22**, pp. 585–586

8 WOOD, P. J.: 'Spherical waves in antenna problems', *Marconi Rev.*, 1971, 34, pp. 149–172

9 SIMMONS, A. J., and KAY, A. F.: 'The scalar feed – a high performance feed for large parabolic reflectors' *in* IEE Conf. Publ. 21 'Design and construction of large steerable aerials', 1966, pp. 213–217

10 LUDWIG, A. C.: 'Radiation pattern synthesis for circular aperture horn antennas', *IEEE Trans.*, 1966, **AP-14**, pp. 434–440

11 VU, T. B., and VU, Q. H.: 'Optimum feed for large radiotelescope-experimental results, *Electron. Lett.*, 1970, **6**, pp. 159–160

12 LUDWIG, A. C.: 'The definition of cross-polarisation', *IEEE Trans.*, 1973, **AP-21**, pp. 116–119
13 RUZE, J.: 'Lateral-feed displacement in a paraboloid', *ibid.*, 1965, **AP-13**, pp. 660–665
14 INGERSON, P. G., and WONG, W. C.: 'Focal region characteristics of offset fed reflectors', *IEEE P. Gap.*, 1974, pp. 121–123
15 THOMAS, B. M.: 'On symmetrical corrugated reflectors with zero depolarisation', *Electron. Lett.*, 1978, **14**, pp. 210–211
16 WOOD, P. J.: 'Tracking sensitivity with peripheral-horn autotrack', *ibid.*, 1973, **9**, pp. 484–485
17 BARTLETT, H. E. et al.,: 'Five horn Cassegrain feed with sidelobe cross-over'. IEEE AP-S Int. Symp., 1977, pp. 332–335

Chapter 9

V.S.W.R. of reflector systems

9.1 Feed-reflector interaction

In any reflector antenna, a portion of the wave scattered off the reflector(s) propagates back into the feed horn and evidences itself as a reflected wave at the antenna input terminals. Unlike the other sources of mismatch arising in a feed system, reflector V.S.W.R. can rarely be compensated over a useful bandwidth by installing a reactive matching element in the waveguide or coaxial feeder line.

Reflector V.S.W.R. may range in magnitude between 1·05 and 1·5 according to antenna geometry. In some applications it must be treated as a feature which is critically interrelated to overall system performance. For example, in a point-to-point microwave link, frequency sensitivity of the input V.S.W.R. can cause distortion for wide-band modulated signals. Another instance which may be cited concerns a circularly polarised antenna for frequency re-use communications. Here an orthomode transducer and polariser are used to excite (say) a RHCP wave in the feed horn. The wave back-scattered from the reflector will be LHCP. Assuming the LHCP port of the OMT is imperfectly matched, there will be a doubly reflected signal which will degrade the isolation between this port and the RHCP port.

Compensation for reflector mismatch ordinarily takes the form of an additional scatterer located near the reflector surface. Some of the principles involved in designing such a device will be described in the following sections.

9.2 Calculation of reflector V.S.W.R.

9.2.1 Field correlation formulation

The value of the reflection coefficient Γ can be specified in terms of the transmitting field of the feed horn incident on the surface of the reflector. According to the field correlation theorem of Section 4.2.2, the voltage multiplier of the component of scattered field of the form $\boldsymbol{E}_t : -\overset{*}{\boldsymbol{H}}_t$, generated in consequence of a field $\boldsymbol{E}_r : \boldsymbol{H}_r$ impinging on the reflector, is

$$\int_S \boldsymbol{H}_r \times \boldsymbol{E}_t \cdot d\boldsymbol{S} \tag{9.1}$$

In the original application of the theorem as cited in Chapter 4, $\boldsymbol{E}_r : \boldsymbol{H}_r$ represented a field associated with a plane wave incident on an antenna operated under receiving conditions. However, for the present purposes, $\boldsymbol{E}_r : \boldsymbol{H}_r$ is to be identified with the incident transmitting field $\boldsymbol{E}_t : \boldsymbol{H}_t$. If the latter distribution is in a form normalised to unit power, the voltage coefficient of eqn. 1 becomes simply the dominant mode reflection coefficient of the feed horn, in respect of reflector interaction. Thus, in general, the electric vector reflection coefficient is given by

$$\Gamma = \frac{-\int_S \boldsymbol{E}_t \times \boldsymbol{H}_t \cdot d\boldsymbol{S}}{R_e \int_{S'} \boldsymbol{E}_t \times \overset{*}{\boldsymbol{H}}_t \cdot d\boldsymbol{S}} \tag{9.2}$$

Here S is the active reflecting surface, and S' is an arbitrary closed surface surrounding the feed horn.

If required, eqn. 2 may be used to evaluate the absolute phase of Γ as well as its amplitude. In this case, Γ is to be calculated at a reference plane (or reference curved surface) at which the electric fields are cophasal (and $\boldsymbol{E} = \overset{*}{\boldsymbol{E}}$). The appropriate absolute phase distributions are assigned to $\boldsymbol{E}_t$ and $\boldsymbol{H}_t$ for an 0° phase electric field at the reference plane. Then after due allowance for the minus sign in the numerator, $\angle\Gamma$ will be correctly evaluated.

Formulations similar to eqn. 2 have been quoted by Poulton[1] (vector version) and Silver[2] (approximate scalar version). For practical feeds the phases of $\boldsymbol{E}_t$ and $\boldsymbol{H}_t$ will be closely correlated across the reflector surface. Thus, the elemental integrand values will add approximately coherently for the denominator integration. However, the values of the numerator integrand tend to combine to generate a spiral phasor diagram, similar to other Fresnel zone phenomena in classical diffraction theory.

The latter feature suggests the possibility of introducing stationary phase principles to avoid the necessity of a numerical integration. Thus, for a front-fed paraboloid there will be a stationary phase point at the vertex,[2] yielding the result

$$\Gamma_{sp} = -j \frac{G_0 \exp(2jkF)}{2kF} \tag{9.3}$$

for the scalar case. However, some approximation is inherent in this simple formula. Thus, for example, with a 15 λ-diameter paraboloid of $F/D = 0{\cdot}433$ and an idealised feed horn model corresponding to a cos θ pattern, truncated at the rim of the dish,

$$\Gamma = 0{\cdot}090 \angle -75{\cdot}75^\circ \qquad \text{(from eqn. 2)}$$

and

$$\Gamma_{sp} = 0{\cdot}084 \angle -86{\cdot}54^\circ \qquad \text{(from eqn. 3)}$$

The phase angle discrepancy between the latter two results could be of some significance when designing a matching system. Moreover, the numerical integrations involved for eqn. 2 are frequently within the scope of a programmable calculator. These considerations should be duly weighed against the attractions of an explicit formula.

9.2.2 *Implementation of correlation formula*

Eqn. 2 requires knowledge of both $\boldsymbol{E}$ and $\boldsymbol{H}$ for the feed horn, at the surface of the reflector. In general, accurate and sufficiently complete information will be available only where a spherical wave expansion has been fitted to extensive measured feed pattern data, or a viable and accurate theoretical model has been established. As was indicated in Chapter 5, the conical corrugated and smooth wall TE_{11}/TM_{11} horns are conducive to analytical modelling.

Assuming a feed horn of the latter type and a coaxially aligned circularly symmetric reflector (reflector V.S.W.R. calculations are unlikely to be required for offset reflectors, since by their design they do not usually reflect significant energy back into the feed), the magnetic field at the reflector surface takes the form

$$\boldsymbol{H}_t = f_1(\theta) \sin \phi \boldsymbol{a}_\theta + f_2(\theta) \cos \phi \boldsymbol{a}_\phi + f_3(\theta) \sin \phi \boldsymbol{a}_r \tag{9.4}$$

Under these conditions the electric field at the surface is given to a very good approximation by

$$\boldsymbol{E}_t = \exp(j\alpha)\{f_2(\theta) \cos \phi \boldsymbol{a}_\theta - f_1(\theta) \sin \phi \boldsymbol{a}_\phi + f_3(\theta) \cos \phi \boldsymbol{a}_r\} \tag{9.5}$$

Eqn. 5 has been deduced from many calculations in which a conical horn model is analysed by S.W.E. techniques, and appears to be quite accurate for all ordinarily encountered Cassegrainian geometries, e.g. ranges of $>0{\cdot}1 D^2/\lambda$ in relation to the feed aperture D. The quantity α is a small fixed phase angle which often does not need to be calculated since it does not affect the magnitude of Γ.

Substituting eqns. 4 and 5 in eqn. 2,

$$\Gamma = -\int_0^{2\pi} \int_0^{\Theta} EH\, r^2 \sin\theta \sec(\theta + \theta_{norm}) d\theta d\phi \qquad (9.6)$$

where (neglecting exp $(j\alpha)$)

$$EH = \{f_1^2(\theta) + f_2^2(\theta)\} \cos(\theta + \theta_{norm}) + \{f_1(\theta) + f_2(\theta)\} f_3(\theta) \sin(\theta + \theta_{norm}) \qquad (9.7)$$

For some applications a slightly more approximate form of eqn. 7 has proved useful. This is

$$EH \simeq f_2^2(\theta) \cos(\theta + \theta_{norm}) + \{f_1(\theta)\cos(\theta + \theta_{norm}) + f_3(\theta)\sin(\theta + \theta_{norm})\}^2 \sec(\theta + \theta_{norm}) \qquad (9.8)$$

Eqn. 8 differs from eqn. 7 only by terms involving $f_3^2(\theta)$ and $\{f_1(\theta) - f_2(\theta)\} f_3(\theta)$ which are usually small. The advantage of this alternative formulation is that it requires only field components parallel to the surface of the reflector (the term in the { } brackets constitutes a tangential field component). Thus eqn. 8 is compatible with the requirements in respect of feed pattern data for conventional physical optics reflector scattering calculations.

In other applications the best available data for the radiation pattern of the feed is a measurement in the far field, and radial field component data (e.g. $f_3(\theta)$) is not available. In this case eqns. 6 and 7 reduce to simply

$$\Gamma \simeq -\int_0^{2\pi} \int_0^{\Theta} \{f_1^2(\theta) + f_2^2(\theta)\} r^2 \sin\theta\, d\theta d\phi \qquad (9.9)$$

where f_1 and f_2 are the measured H- and E-plane patterns corrected for space attenuation at the reflector surface and normalised to unit power,

$$\int_0^{2\pi} \int_0^{\pi} \{|f_1(\theta)|^2 + |f_2(\theta)|^2\} r^2 \sin\theta\, d\theta d\phi = 1 \qquad (9.10)$$

9.2.3 Role of transmitting focal plane distribution

The field re-impinging on the plane of the feed aperture after being scattered by the reflector constitutes a 'transmitting focal plane distribution'. It can conveniently be calculated by a Taylor spherical wave series in a manner similar to the receiving focal plane case of Chapter 8. The only distinction is that the starting point is now the currents induced by transmitting fields launched by the feed horn, rather than those induced by a received plane wavefront.

An example of a transmitting FPD, for the case of a 10λ $F/D = 0{\cdot}4$ paraboloid, is illustrated in Fig. 9.1.

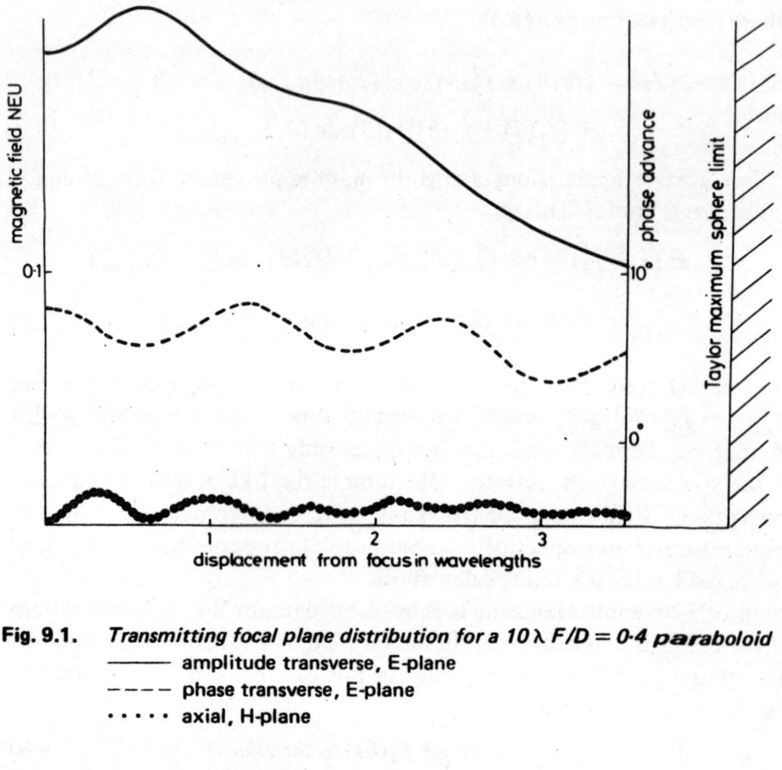

Fig. 9.1. *Transmitting focal plane distribution for a 10λ $F/D = 0{\cdot}4$ paraboloid*
——— amplitude transverse, E-plane
– – – – phase transverse, E-plane
• • • • • axial, H-plane

9.3 A reflector matching theorem

9.3.1 Definitions

We define the nominal reflection coefficient Γ_{nom} as that for the complete nominal, unmatched reflector surface. The local reflection coefficient Γ_l is similarly defined, but in relation to an elemental cell of the reflector surface. For example, if this cell is assumed to take the form of an annulus specified by θ limits θ_A and θ'_A, then Γ_l could be evaluated by substituting these limits in an integral formulation such as eqn. 6. When a matching device is superimposed on the nominal reflecting surface, the difference between reflection coefficients with and without matching may be represented as a perturbation reflection coefficient Γ_p.

9.3.2 Constraints for Γ_p

Suppose that the matching device is to consist of a perturbation to the nominal reflector surface which does not destroy its inherent 'single valued' property as viewed from the feed (strictly, this stipulation excludes devices such as the spaced vertex plate which nominally shadow an area of the reflector, but may nevertheless allow some induced current). Now consider the scattering from a small cell of the perturbation region located immediately above a point X on the nominal surface.

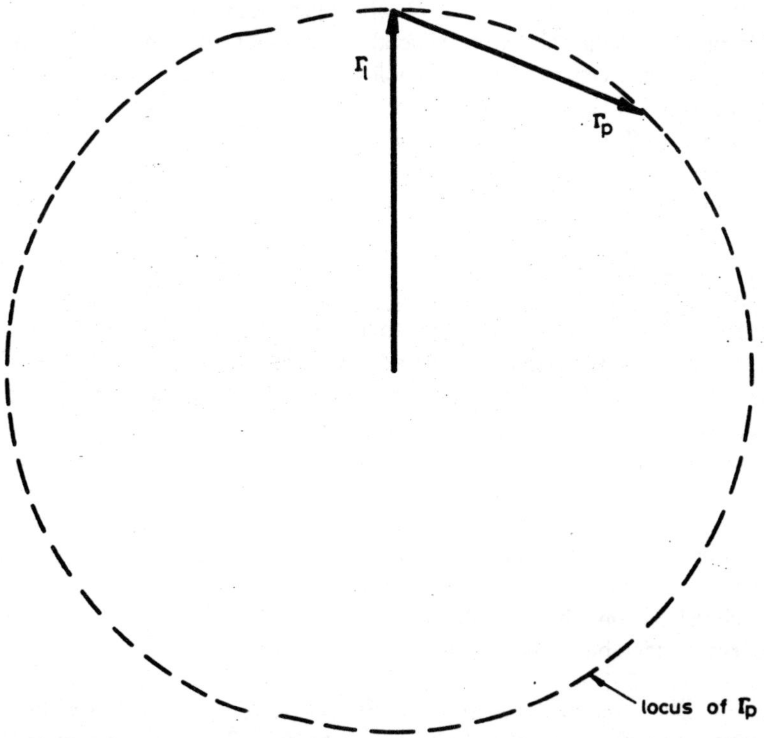

Fig. 9.2. *Fundamental limitations to matching by a small profile perturbation*

As is illustrated by Fig. 9.2, $\angle\Gamma_p$ for this cell is partly fixed by $\angle\Gamma_l$ for the nominal surface at X. In fact, the degree of freedom available amounts to an angle whose modulus certainly cannot exceed $\pi/2$. In practice, $\angle\Gamma_p$ must duplicate $\angle -\Gamma_l$ within about $\pm\pi/4$, if the perturbing device is to be reasonably efficient in generating an additional reflection.

The significance of this theorem can readily be appreciated in the context of a simple paraboloid antenna. To clarify the discussion, suppose that the feed horn takes the form of an essentially cophasally illuminated plane aperture, located a large integral number of wavelengths from the reflector vertex. Then Γ_{nom} will have a phase angle of the order of $-\pi/2$ at the aperture plane, from eqn. 3 or eqn. 6.

$\angle\Gamma_l$ will be $0°$ at the vertex of the dish. Conversely, at a point on the nominal reflecting surface just sufficiently displaced from the vertex to increase the electrical path length to the feed by $\pi/4, \angle\Gamma_l = -\pi/2$. Therefore, by perturbing the surface near this latter point, a Γ_p having a phase angle in the range $\pi/4$ to $3\pi/4$ can be generated.

It may be deduced that in general there will be a hollow annulus region, near the centre of the paraboloid and bounded by two coaxial circles, which is capable of generating an appropriate $\angle\Gamma_p$ for compensating Γ_{nom}. The region within the inner annulus (encompassing the paraboloid vertex) ordinarily makes only a second order contribution to Γ. Thus the annulus may be extended into this region without significant degradation of performance. In the limit the annulus degenerates into a simple circular disc, popularly known as a 'vertex plate'.[2]

It cannot be over-emphasised that the viability of vertex plate matching depends upon the phase relationship of Γ_{nom} and Γ_l for the nominal profile. For antennas other than a paraboloid it is often necessary to reintroduce the annulus-type geometry (e.g. position the matching device away from the vertex), to achieve a match efficiently. 'Efficiently' here means in a manner such that the scattered field from different parts of the matching device add essentially coherently at the feed horn. A lack of coherency would imply an excessively large profile perturbation to achieve compensation, with attendant undesirable repercussions on the overall antenna performance, particularly in respect of sidelobe levels.

9.4 Matching by quadrature profile perturbation

9.4.1 Use of physical optics principle

The vertex plate matching technique has enjoyed some popularity because it uses an 'add-on' device which readily lends itself to empirical 'cut-and-try' optimisation on the test bench. By the same token, it is often not feasible to apply analytical optimisation methods. Thus, the physical optics principle of Section 2.2.6 is ordinarily applied to a reflecting surface of essentially smooth profile. However, a PO treatment may fail in the case of a vertex plate, either because of the small diameter in relation to the wavelength, or (more usually) because there is a surface discontinuity or gap at the periphery of the plate.

The absence of accurate analytical design procedures makes the vertex plate solution less attractive when antenna specifications are particularly stringent. Moreover, vertex plate matching involves some degradation in overall antenna performance characteristics, particularly in respect of gain and sidelobes. There are also fundamental limits on the match bandwidth that can be achieved.

The alternative to 'add-on' devices is a 'quadrature profile perturbation'. The reflector profile is slightly distorted so as to introduce a Γ_p which is approximately in phase quadrature with Γ_l at each point on the profile. From the manufacturing point of view, this approach is particularly attractive for a shaped Cassegrainian subreflector. In this type of application, a numerically-controlled machine would be used to make a mould or blank even for an unmatched reflector. The incorporation of matching merely involves a change of co-ordinates on the control tape.

The quadrature profile perturbation method allows continuity of the reflector surface, and its first derivative, to be preserved. Thus physical optics design methods can be implemented.

9.4.2 Continuous quadrature profile perturbation

Ideally, a perfect match could be achieved if the transmitting focal plane distribution were to incorporate a deep null encompassing the feed horn aperture. To avoid perturbing the overall antenna aperture distribution and its far field radiation properties, the null should be nearly rectangular, with steep sides (e.g. see Fig. 7.14). This in turn implies that the whole of the reflecting surface should be perturbed to synthesise an efficient transmitting FPD.

In the Cassegrainian case, a form of approximation to these conditions can be brought about as an incidental feature of the diffraction profile synthesis procedure described in Section 7.3. When a feed system shadow is introduced into the DPS model, the subreflector profile generated is such as to cause scattered power to be diverted from the central region of the main reflector. That is, there is a tendency to focus a shadow at the range of the main reflector whose width roughly matches the feed system blockage.

At the feed horn aperture, however, this shadow is partly blurred. A finite null depth is to be expected in any event, in that the criterion for DPS synthesis is maximum overall far field efficiency rather than minimum power in nominally shadowed near field regions. Moreover the optimisation process makes no distinction between power which is refocused into the dominant waveguide mode of the horn, and re-scattered power.

For all these reasons, DPS synthesis can in itself only provide a modest reduction in Γ, perhaps 6 dB. Compensation is broadband (e.g. 50%), but inevitably the sides of the synthesis null spread beyond the feed system shadow region. In a typical case this spreading effect is equivalent to a 50% increase relative to the geometrical shadow diameter. As a consequence, the far field blockage sidelobe structure is substantially augmented.

9.4.3 Discrete quadrature profile perturbation

Consider a profile distortion defined by

$$\Delta r = \tfrac{1}{2} U(1 + \cos \pi x), \qquad |x| < 1 \tag{9.11}$$

$$\Delta r = 0, \qquad |x| > 1 \tag{9.12}$$

where

$$x = \frac{2(\theta - \theta_a)}{\theta_b - \theta_a} - 1 \tag{9.13}$$

and U and $\theta_b - \theta_a$ are small quantities (Fig. 9.3). Eqns. 11–13 represent a raised cosine perturbation of the annulus defined by angles θ_a and θ_b.

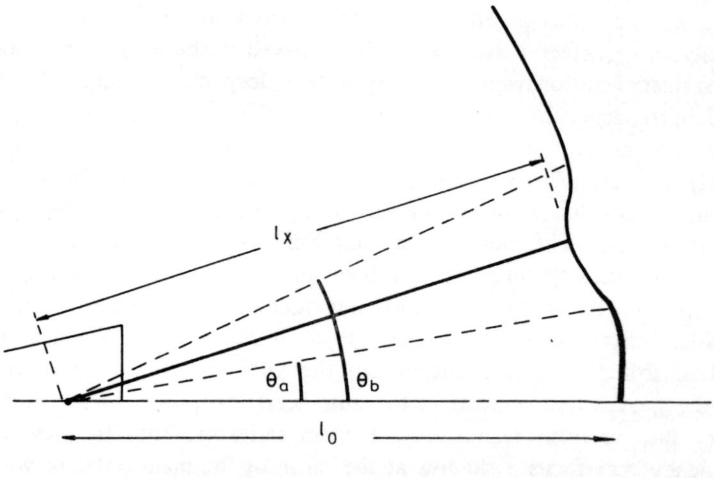

Fig. 9.3. *Geometrical parameters for a raised cosine annulus*

Such a discrete annulus will inevitably modify the transmitting FPD over the whole focal plane of the antenna, as opposed to merely creating a null in the central region as discussed in Section 9.4.2. However, it is possible to provide a combination of inphase and antiphase excitation of the horn aperture that involves no net coupling to the dominant waveguide mode. Higher-order modes will of course be excited and re-radiated as wide angle sidelobes.

For a single matching annulus, the parameters of eqns. 11–13 may be optimised by a trial and error method. For any particular trial, Γ_p may be evaluated as the complex difference of Γ values calculated according to Section 9.2.2, for the nominal profile, and for the perturbed profile. The parameters for the next trial may then be selected on the basis that

$$\angle\Gamma_p \simeq \angle\Gamma_l + C_2$$

$$\simeq -2kl_X + C_3 \tag{9.14}$$

and

$$|\Gamma_p| \simeq C_1 U(\theta_b - \theta_a) \tag{9.15}$$

Here C_1, C_2, C_3 are constants. C_2 will be of the order of $-\pi/2$ for a depression perturbation, or $\pi/2$ for a raised perturbation, according to Section 9.3.2. The parameter l_X is the geometrical path length from the midpoint

$$\theta_{an} = \tfrac{1}{2}(\theta_b + \theta_a)$$

of the annulus to the feed horn.

For a specified $\angle\Gamma_p$, eqn. 14 yields many possible solutions for θ_{an}. However, the smallest solution, representing an annulus of minimum possible diameter, will ordinarily be selected in the interests of V.S.W.R. bandwidth.

For a specified $|\Gamma_p|$, eqn. 15 defines only the product $U(\theta_b - \theta_a)$. In this case, it is expedient to allow $\theta_b - \theta_a$ to take on the largest value that is in keeping with a coherent reflection from all parts of the annulus, e.g.

$$k(r_b - r_a) = \pi/2 \tag{9.16}$$

Usually, conformance to this relationship will ensure that U is sufficiently small to avoid either manufacturing difficulties associated with excessive profile curvature or analytical problems due to failure of the physical optics principle.

An optimisation process based on the principles described above ideally will result in a 1:1 V.S.W.R. at design frequency f_0. However, the match will degrade rapidly away from this frequency. The bandwidth performance again can be evaluated using eqn. 6. A useful approximate rule in this context is that $\angle\Gamma_p$ will vary relative to $\angle -\Gamma_{nom}$ as twice the difference between two electrical path lengths. These correspond to the geometrical lengths l_X and l_0 of Fig. 9.3.

Optimum raised cosine annulus parameters for a particular 13 λ diameter shaped Cassegrainian subreflector were

$$U = 0{\cdot}095\,\lambda$$

$$\theta_a = 5{\cdot}72^\circ$$

$$\theta_b = 9{\cdot}52^\circ$$

The unmatched profile has a Γ_{nom} of 0·097. The transmitting FPD, at the design frequency f_0 at which the overall match is nearly perfect, is given in Fig. 9.4.

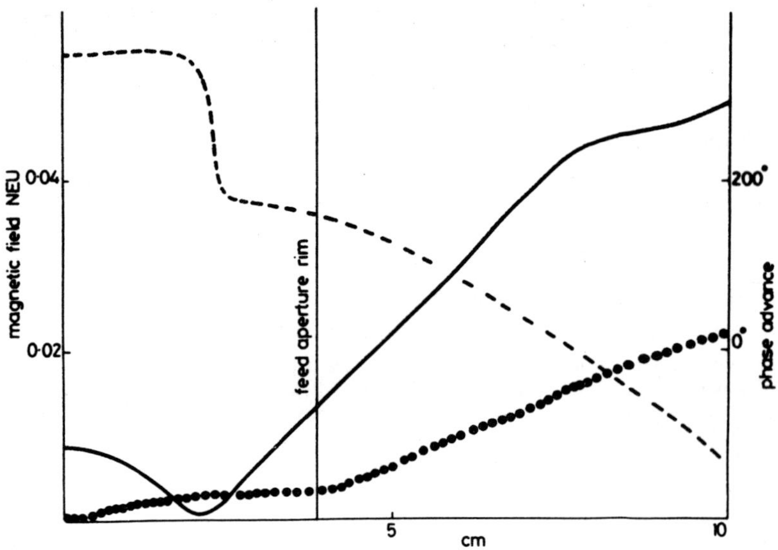

Fig. 9.4. *Transmitting FPD for single annulus matched feed system*
——— amplitude transverse, E-plane
– – – – phase transverse, E-plane
• • • axial, H-plane

The MARECS shaped beam earth coverage antenna application was introduced in Section 3.4.5. The MARECS antenna launches nominal right hand circular polarisation. That part of an orthogonally polarised scattered field which is captured by the feed horn tends to be harmlessly absorbed in a matched load port of the polarising hybrid. However, any power rescattered from the feed horn suffers a second polarisation inversion, and hence could cause frequency sensitive changes in the shaped beam pattern. This leads to a requirement for a matching system which will reduce the overall level of near field energy in the vicinity of the horn. The profile perturbation used[3] is an adaptation of a form previously proposed[4] for paraboloid antennas. Its performance is illustrated by the transmitting FPD of Fig. 9.5.

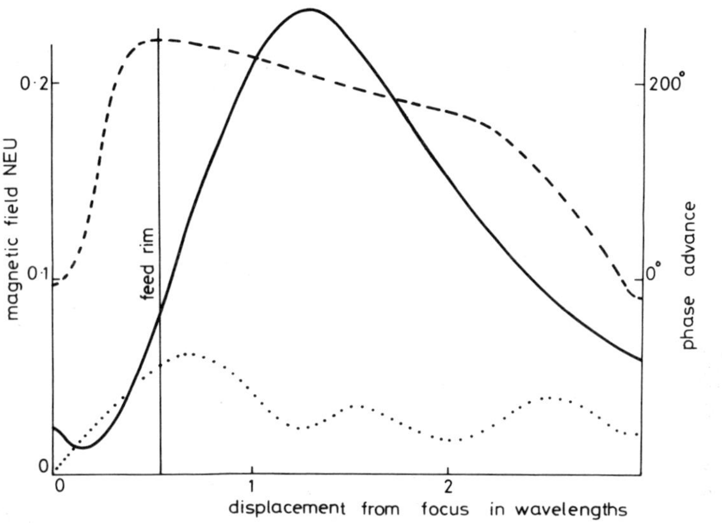

Fig. 9.5. *Transmitting FPD for a front-fed shaped beam reflector matched by profile perturbation*

——— amplitude transverse E-plane

– – – – phase transverse E-plane

· · · · · axial H-plane

9.5 A matching system for two frequency bands*

As an example of the versatility of the quadrature profile perturbation technique, a circularly polarised Cassegrainian feed system will be described that is matched over two frequency bands, 11·45 to 11·8 GHz, and 14 to 14·5 GHz.[5] Initial Cassegrain profiles were calculated by the DPS technique of Chapter 7. The feed system shadow was not taken into account in the model during synthesis iterations, and the nominal profiles are essentially unmatched as for the case of Section 9.4.3. Γ_{nom} is 0·097 at the 11·5 GHz and 0·105 at the 14·45 GHz design frequency, e.g. 1·23:1 V.S.W.R.

The matching element for the subreflector is of the cosine annulus

* Patent application pending

type previously described. However, two annuli are now used (Fig. 9.6). The aim is to achieve a condition whereby $\angle\Gamma_{p2}$ for the outer (correction) annulus varies from approximately lagging quadrature to leading quadrature relative to $\angle\Gamma_{p1}$, as the frequency is varied from design frequency f_1 to design frequency f_2. In this way it is possible to

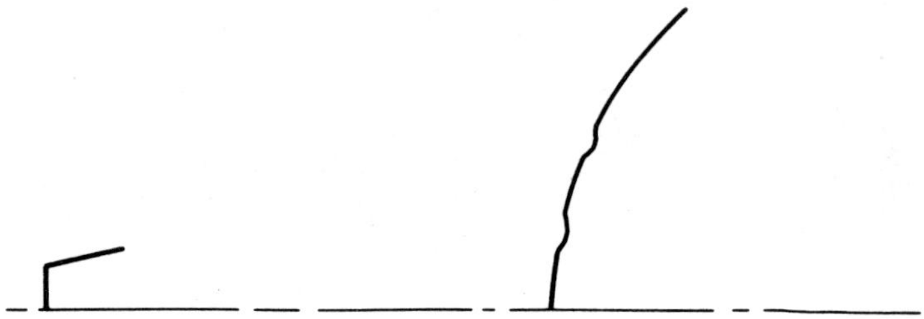

Fig. 9.6. *Subreflector with two matching annuli*

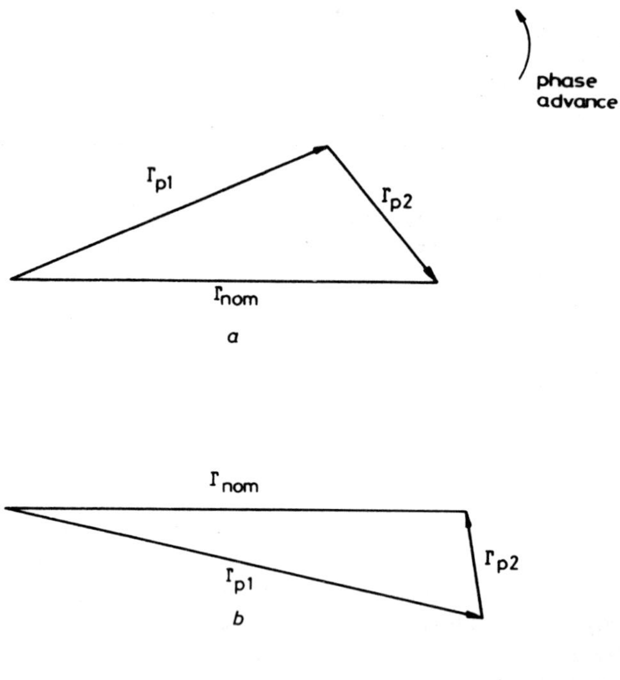

Fig. 9.7. *Phasor diagrams illustrating principles of a dual band matched feed system*

a 11·5 GHz

b 14·45 GHz

introduce a correction for the variation relative to Γ_{nom} of the inner annulus $\angle \Gamma_{p1}$.

The geometrical parameters for each annulus were determined by trial and error evaluations of eqn. 8, in the manner described in

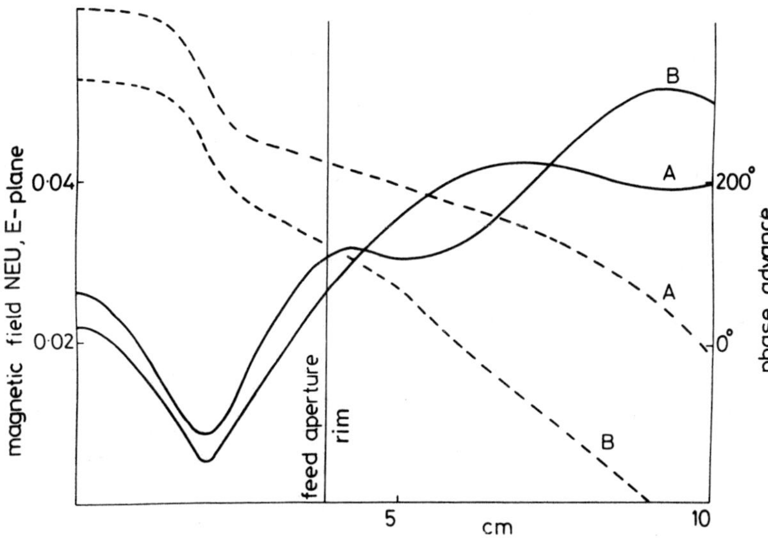

Fig. 9.8. *Transmitting FPD for dual band feed system*
——— amplitude
– – – – phase
A 11·50 GHz
B 14·45 GHz

Section 9.4.3. The following solution was obtained.

$$U_1 = 0{\cdot}079\,\lambda$$

$$\theta_{a1} = 5{\cdot}4^\circ$$

$$\theta_{b1} = 9{\cdot}2^\circ$$

$$U_2 = 0{\cdot}092\,\lambda$$

$$\theta_{a2} = 17{\cdot}22^\circ$$

$$\theta_{b2} = 20{\cdot}22^\circ$$

The reflection coefficient phasor diagrams of Fig. 9.7 indicate how Γ_{p1} and Γ_{p2} combine to provide a match at two discrete frequencies. Fig. 9.8 demonstrates how the transmitting FPD decouples the dominant mode of the horn by providing appropriate inphase/antiphase excitation of the aperture at each frequency.

Fig. 9.9 illustrates the assembled 3 m Cassegrain feed system. The shaped subreflector V.S.W.R. has been calculated as a function of frequency from eqn. 8, and also measured. The experimental technique

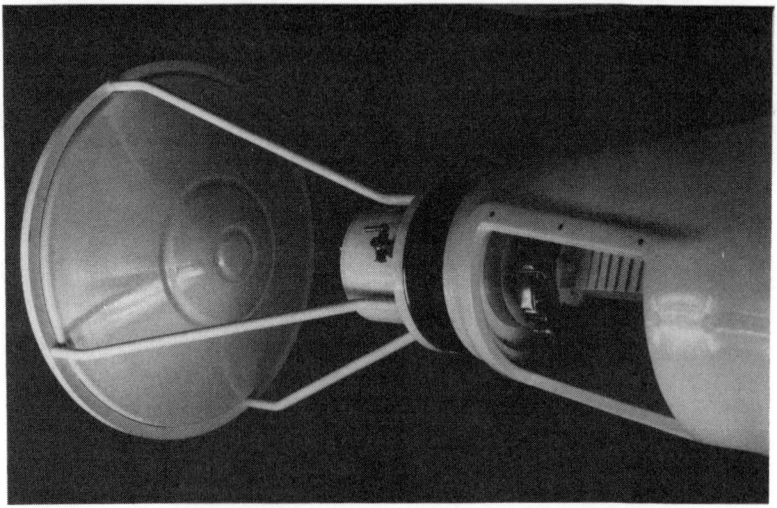

Fig. 9.9. *Dual frequency matched feed system*

involved a measurement of complex Γ in the feeder waveguide with and without an absorber sheet shielding the subreflector. By subtracting the two measured values, the effect of waveguide component mismatches was eliminated. As can be seen from Fig. 9.10, the experimental results are in good agreement with the theoretical prediction. Such small discrepancies as are evident are believed to be due to multiple reflection effects not taken into account in the theory.

The dual annulus matching device introduces some phase error in the subreflector scatter pattern. However, this may be compensated by suitably perturbing the surface of the main reflector. Thus, the original main reflector profile, as generated by the DPS technique to suit the nominal (unmatched) subreflector, is now discarded. Instead, the synthesis process is repeated in a 'constrained subreflector' role (see Section 7.6.1), taking into account the new matched subreflector profile. Thus, main reflector co-ordinates are generated which are gain-matched to the feed system.[5] The overall result is that both the gain and far-field radiation pattern performance (including sidelobes) of the final reflector configuration are almost identical to that predicted for nominal unmatched profiles. The theoretical prediction in the latter case was based on the somewhat unrealistic hypothesis that the wave reflected back into the feed horn would be absorbed without any detriment to gain or radiation pattern.

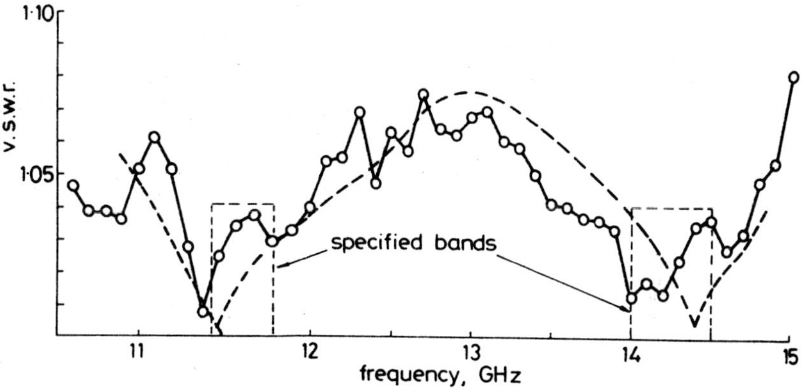

Fig. 9.10. *V.S.W.R. for the dual band feed system*
–o–o– empirical
– – – – theory

References

1 POULTON, G. T., and LIM, S. H.: 'Calculation of input-voltage standing-wave ratio for a reflector antenna', *Electron. Lett.*, 1972, **8**, pp. 610–611

2 SILVER, S.: 'Microwave antenna theory and design', (McGraw-Hill, 1949, pp. 155–158

3 WOOD, P. J.: 'Shaped beam reflector design for earth coverage applications'. IEE Antennas for Aircraft and Spacecraft Conference, London, 1975, pp. 47–52

4 POULTON, G. T. and ALMOAYYED, T.: 'Optimum vertex plate design for reflector antennas'. European Microwave Conference 1973, paper C.4.4

5 BRAIN, J. R., and WOOD, P. J.: 'Problems inherent in the design and measurement of an 11/14 GHz antenna' IEE International Conference on Antennas and Propagation, Nov. 1978, pp. 213–217

Index